Praise for *Tractor Wars*

"The mechanization of agriculture revolutionized American farm production in the late 1800s and early 1900s. Neil Dahlstrom's *Tractor Wars* examines an important facet of this historical watershed—the development of the farm tractor. Exploring the complex relationships among industrialist Henry Ford, Alexander Legge of International Harvester, and William Butterworth of John Deere—the three key players in bringing the tractor to thousands of farmers—this book tells a fascinating story of innovation and intrigue, competition and contention, alliances and animosities that produced this essential agricultural machine. Dahlstrom has dug deeply into archival sources and writes with clarity and insight, and his *Tractor Wars* provides a valuable book for the historian and a compelling one for the general reader interested in American rural life."

—Steven Watts, author of *The People's Tycoon: Henry Ford and the American Century*

"Neil Dahlstrom's *Tractor Wars* engagingly tells the story of one of the great business battles of the twentieth century. Facing the power of Henry Ford, who had long wanted to build a better tractor, and the International Harvester Company, a giant combine formed through J.P. Morgan, the smaller John Deere Company found a way to compete in the emerging tractor business. Anyone interested in business, agriculture, or tractor history will enjoy this great tale, well-told."

—Gary Hoover, Executive Director, American Business History Center

"For all the attention given to moments in technological innovation brought about by events like the race to the moon, it's easy to overlook a period that was arguably as transformational a hundred years ago—one that happened 'down on the farm.' In *Tractor Wars*, author and industry expert Neil Dahlstrom brings his own deep understanding of business and history to harvest a deeply compelling story about the colorful characters behind a two-decade battle that would affect everything—from the food we put on the table—to the growth of our cities and, ultimately, an American landscape fundamentally transformed. *Tractor Wars* breaks new ground in the American narrative of the twentieth century. Better still for readers, Dahlstrom's journey by tractor is a genuinely delightful ride."

—David Brown, host of the *Business Wars* podcast and author of *The Art of Business Wars*

TRACTOR WARS

Also by Neil Dahlstrom

The John Deere Story: A Biography of Plowmakers John and Charles Deere

Lincoln's Wrath: Fierce Mobs, Brilliant Scoundrels, and a President's Mission to Destroy the Press

TRACTOR WARS

John Deere, Henry Ford, International Harvester, and the Birth of Modern Agriculture

Neil Dahlstrom

Matt Holt Books
An Imprint of BenBella Books, Inc.
Dallas, TX

Tractor Wars copyright © 2022 by Neil Dahlstrom
First trade paperback edition 2025

Except in the case of brief quotations embodied in critical articles or reviews, no part of this book may be used or reproduced, stored, transmitted, or used in any manner whatsoever, including for training artificial intelligence (AI) technologies or for automated text and data mining, without prior written permission from the publisher.

Matt Holt Books is an imprint of BenBella Books, Inc.
8080 N. Central Expressway
Suite 1700
Dallas, TX 75206
benbellabooks.com
Send feedback to feedback@benbellabooks.com.

BenBella and *Matt Holt* are federally registered trademarks.

Printed in the United States of America
10 9 8 7 6 5 4 3 2 1

Library of Congress Control Number: 2021034770
ISBN 9781637744987 (trade paper)
ISBN 9781953295743 (trade cloth)
ISBN 9781637740088 (ebook)

Editing by Katie Dickman and Brian Nicol
Copyediting by Rebecca Taff
Proofreading by Jenny Bridges and Marissa Wold Uhrina
Indexing by Debra Bowman
Text design and composition by PerfecType, Nashville, TN
Cover design by Kara Klontz
Cover photos provided courtesy of John Deere (center image only) and the Wisconsin Historical Society; frame © Shutterstock / K3Star
Printed by Lake Book Manufacturing

Special discounts for bulk sales are available. Please contact bulkorders@benbellabooks.com.

To Karen and Grant, for giving me early mornings, late nights, and long weekends, and for entertaining endless tractor stories.

CONTENTS

KEY TRACTORS AND PERSONNEL

Key Tractors (and year of first introduction)	Key Personnel
Ford Motor Company/Henry Ford & Son	
M.O.M. (Ministry of Munitions) (1917)	Henry Ford
Fordson (1918)	Edsel Ford
	Eugene Farkas
	Joseph Galamb
	Charles Sorensen
	Lord Percival Perry
John Deere	
Melvin (experimental)	William Butterworth
Sklovsky (experimental)	Willard Velie
Tractivator (experimental)	George Mixter
Waterloo Boy (1914)	Theo Brown
John Deere Tractor (Dain) (1918)	Joseph Dain, Jr.
D (1923)	Leon Clausen
GP (aka All-Crop, C, Power-Farmer) (1928)	Charles Deere Wiman

Key Tractors (and year of first introduction)	Key Personnel
International Harvester	
Mogul (1909)	Cyrus McCormick, Jr.
Titan (1916)	Alexander Legge
Motor Cultivator (1918)	A.E. Johnston
McCormick-Deering 15-30 (1921)	Bert Benjamin
McCormick-Deering 10-20 (1923)	
Farmall (1923)	

PROLOGUE

Working from sunup to sundown in the mild cold, the harsh wind, the driving rain, or whatever weather surprises *Old Farmer's Almanack* tried its best to predict, Irish immigrant William Ford and his family worked to finish the harvest on their modest acreage in Dearborn, Michigan, outside of Detroit. The scene was repeated by millions of families each year across the country.

The farm offered an honest living to those willing to work for it. The consequences of each day were on exhibit at each meal, often taken in the field. Success was marked daily by a well-nourished, or angrily empty, stomach.

Henry Ford, William's son, was twelve years old when a new sound captured his imagination. In 1875, the young man and his father crossed paths with a curious contraption on four wheels, sitting lifeless on the long-worn dirt road it had been traveling just outside of Detroit. Writing nearly fifty years later, Ford remembered "that engine as though I had seen it only yesterday, for it was the first vehicle other than horse-drawn that I had ever seen."

The steam engine, which he learned was made by Nichols, Shepard & Company of Battle Creek, Michigan, had a chain drive between the engine and its two rear wheels. "It was intended to drive threshing machines and power sawmills and was simply a portable engine and a boiler mounted on wheels," Ford later described, reminiscing about his long, inquisitive conversation with the engineer.

Henry Ford, a young man with demonstrated mechanical aptitude, limitless imagination, and ambition, left the farm at the age of sixteen for the city of Detroit. But in reality, he never really left. Nor could he escape the memories, sounds, and cycles of the farm, or what confounded him the most, the too easily accepted traditions of inefficiency and "drudgery," as he called it, that kept American farmers from realizing their potential.

In time, as Americans developed an appetite for the automobile, Henry Ford directed his talents toward its advancement. The Model T debuted in October 1908, irreversibly accelerating American mobility and culture.

Less than a month later, Ford submitted a short note to the editors of the *Farm Implement News*, a leading farm journal published in Chicago. Attached was a photo of himself at the wheel of an experimental vehicle assembled from discarded automobile and farm implement parts. While Americans raced to the nearest Ford dealer to buy the people's car, its creator retreated to the farm to finish the work he had been dreaming about since that fateful encounter at the age of twelve.

He was not alone.

From barns, sheds, machine shops, and factories in Illinois, Minnesota, Michigan, and California, to the endless Canadian prairies and rolling farm fields of North Dakota, the race to revolutionize American agriculture had already begun. The machine was called a tractor.

1

THE FORD TRACTOR

Henry Ford's Model T was, in his mind, perfection. In the decades after its debut, few Americans disagreed. Still, the Model T was only part of Ford's vision for the new American century. Well before he began to think about transportation, the man who by 1913 was building half of the automobiles sold in the United States "had the idea of making some kind of a light steam car that would take the place of horses—more especially, however, as a tractor to attend to the excessively hard labour of ploughing."[1]

Ford's relentless pursuit of the farm tractor was driven by his mechanical abilities, a growing nostalgia for his family's homestead, and a mixed empathy and contempt for farmers, who he thought were unprepared for the technological revolution before them. "From the beginning, I never could work up much interest in the labour of farming," he wrote. But solving the centuries-old struggle to make the farm efficient was certainly worthy of his attention.

Soon after his first encounter with a steam engine, Ford was crawling across railroad yards to watch the steam-belching locomotives up close, seeking to understand the complex relationship between size and power. As his understanding of mechanical power grew, so too did his disdain for human inefficiency and wasted labor. His sister, Margaret, said he "wanted things done with the least loss of time and energy. If a job could be done more simply that was the way it should be done."[2]

Ford continued to fine-tune his mechanical abilities under the daily direction of his father, William, and benefited from the well-equipped workshop on the family farm. William had a taste for technology as well. In 1876, he was one of the nearly ten million people drawn to the Centennial Exposition, the first World's Fair held in the United States, on the grounds of Fairmont Park in Philadelphia. There, "never before in the history of mankind have the civilized nations contributed such a display of their peculiar treasures," offered a souvenir publication. William enthusiastically watched "scheduled tests of the steam plows and road engines," the portable steam engines of the Ames Iron Works, Oswego, New York, or a Bigelow Stationary Engine of the type installed in growing numbers in factories across the country. Margaret remembered her father and his friends talking about "the mechanical exhibits and the wonderful things that could be accomplished by steam power."[3]

Not long after, Henry found employment as an apprentice at a small machine shop in Detroit, then began working on steam engines at the Detroit Dry Dock Company, steamship builders located at Atwater and Orleans Street. One of its owners would later become one of Ford's earliest investors. In the fall, Henry returned home to help bring in the harvest.[4]

The Fords were a common, self-sustaining farm family. Henry and his siblings fed and cared for the animals, fetched water, helped with the cooking and cleaning, and joined in the monotonous manual labor that defined each day. He grew increasingly impatient with its operation over time. He vividly remembered his father buying a harvester, a horse-drawn implement that cut grain, in 1881 and, years later, as an adult, was bewildered that so little advancement had been made on the basic machine form. Everything was still done by hand, he later wrote, and should be replaced by machinery. "Power-farming is simply taking the burden from flesh and blood and putting it on steel," he rationalized.

"The farmer makes too complex an affair out of his daily work," he penned in *My Life and Work*, an autobiographical manifesto of his methods and motivations written after the world had crowned him the "flivver king," a reference to his cheap automobile. "Power is utilized to the least possible degree. Not only is everything done by hand, but seldom is a thought given to logical arrangement. A farmer doing his chores will walk up and down a rickety ladder a dozen times. He will carry water for years instead of putting in a few lengths of pipe. His whole idea, when there is extra work to do, is to hire extra men. He thinks of putting money into improvements as an expense." To Ford, the waste of farm inefficiency led to wasted effort, high prices, and low profits.[5]

"Nothing could be more inexcusable than the average farmer, his wife, and their children drudging from early morning until late at night," he wrote from years of experience.[6]

Henry, lanky with a short neck and dark, short, curly hair, married Clara Bryant, the oldest of twelve children raised on a farm in nearby Greenfield, in 1888. She was smart and progressive, and Henry's biggest supporter and advocate. In September 1891, they rented half of a two-story house at 618 John R Street in Detroit for ten dollars a month. Ford secured a position with the Edison Illuminating Company, which had begun the work of lighting the streets and homes of America. Ford spent the next eight years there, first working the overnight shift from 6 PM to 6 AM, supervising the generators at the Willis and Woodward substation, which supplied electricity for the houses in the neighborhood. Within three years his ability to troubleshoot issues was well-known across the company.[7]

Ford devoured the pages of popular magazines and journals, and at home, now in his thirties, he spent nights designing and building an internal combustion engine of his own design, an evolution from previous efforts in Dearborn to design a steam-powered tractor. This work convinced him that a gasoline engine could power a horseless carriage. In 1893, he attended the Columbian World Exposition in Chicago and was first captivated by a two-cylinder Daimler engine mounted on a cart. Gottlieb Daimler, a fifty-nine-year-old engineer from Germany, received awards for both of his exhibits, an automobile and a boat propelled with combustion engines. Inspired by all he had seen in Chicago, Ford returned home and set to work mounting his own engine to a bicycle.

It was a life-changing year. Clara gave birth to their first child, Edsel, on November 6, and Henry was promoted to Edison's chief engineer in Detroit. Ironically, a promotion and a raise facilitated more free time. His work varied so much that he found time to tinker, to visit machine shops, even to teach a nighttime metalworking class at the YMCA.[8]

Ford the automaker began to emerge. At their latest residence, a brick double house at 58 Bagley Avenue (the neighbor was number 56), he worked through ideas in the small coal shed out back. Using salvaged parts and tools from work, he constructed a four-cycle engine from a design featured in the pages of a trade magazine, bringing it inside and setting it on the kitchen sink. It was Christmas Eve 1893. With one-month old Edsel in another room, Ford connected a wire to a kitchen light bulb to create ignition, and while Clara fed the engine drips of fuel, Henry turned the small flywheel. "I didn't stop to play with it," he wrote. "I wanted to build a two-cylinder engine . . . and started work on it right away."

⁂

Over the next several years Ford and a growing group of mechanics, engineers, and other associates worked nights and weekends on what eventually became the Quadricycle. The four-cycle, two-cylinder motor, generating three to four horsepower, a calculation of pulling power, was mounted on four twenty-eight-inch bicycle wheels with rubber tires, running at a low speed of ten miles per hour and a high speed of twenty-four miles per hour. It was his first automobile.

On June 4, 1896, Ford set out to test his Quadricycle for the public after literally running into a brick wall. The team had built the vehicle larger than the width of the garage door. Anxiously, Ford grabbed an axe and began knocking out bricks until they could push the vehicle outside. After a minor mechanical setback, Ford turned on the current from the battery, adjusted the fuel, choked the engine, and cranked the flywheel. The automobile started down Grand River Avenue, stalled, and after a quick repair to a spring, was off again.

Ford's development came during a euphoric time in the early history of the automobile. Powered by gasoline, kerosene, or in many cases, electric batteries, automobiles were being built and tested on the streets of Philadelphia, New York, Detroit, Chicago, and smaller towns like Waterloo, Iowa, and Moline, Illinois. But unlike others who failed to improve their designs or rushed to production and sales, Ford immediately sold his creation and put the $200 in proceeds toward his next iteration. "I had built the first car not to sell but only to experiment with. I wanted to start another car."

He soon received much needed encouragement from his hero, Thomas Edison. In August 1896, Ford was invited to a convention of the Edison organization in New York City at the opulent Oriental Hotel. At one point, he found himself sharing dinner with some of the greatest minds of the electrical industry, including Edison himself. Alexander Dow, the brilliant manager of the Edison Illuminating Company, added to a conversation about storage batteries for electric cars, pointing out Ford's own work on a gasoline automobile. Edison, who himself had been developing an electrically powered vehicle, was enthusiastic about Ford's work. Soon, the two were talking and sketching ideas on the back of a menu card.

Ford left the convention with everything he needed to continue his work—encouragement from Edison himself. "Young man, that's the thing; you have it," Edison responded, banging his fist on the table as Ford sketched. "Keep at it! Electric cars must keep near to power stations. The storage battery is too heavy. Steam cars won't do it either, for they have to carry a boiler and fire. Your car is self-contained—it carries its own power plant—no fire, no boiler, no smoke, no steam. You have the thing. Keep at it."

"That bang on the table was worth worlds to me," Ford wrote. "No man up to then had given me any encouragement."

Ford's second car was finished in late 1897, with modifications continuing into 1898. He raised $500 each from four investors, Mayor William Maybury, Ellery I. Garfield of the Fort Wayne Electrical Corporation, Everett A. Leonard of the Standard Life and Accident Insurance Company, and local physician Dr. Benjamin Hoyt Rush. The next prototype drew attention from the media, leading to additional investors with deeper pockets and important business connections.[9]

From his first partnership, Ford found that his vision conflicted with the expectations of the financial backers, who looked only for a financial return. After several false business starts, he formed a partnership with Alexander Malcomson, a self-made Detroit businessman whose financial resources were derived from a massive coal delivery network. They incorporated the Ford Motor Company in June 1903, soon adding additional investors. That summer, assembly of what they called the Model A began in a rented building on the corner of Mack Avenue in Detroit. Over the next year, sales slowly grew, and in April 1904, a larger, three-story factory was purchased on Piquette Avenue. Operations were completely relocated by early 1905, now with a sizable workforce of three hundred people.[10]

Ford still preferred to work in his own shed not far away from the Piquette Avenue factory as their lineup expanded and profits fueled dissension among the board. The company was selling four models, the Model A, and the newly introduced Model C and Model F, which were inexpensive, two-cylinder touring vehicles. A fourth model, the four-cylinder Model B, showed modest sales comparatively.

Two more models were developed for 1905–1906, the four-cylinder Model N and the six-cylinder Model K, which represented a growing rift between Ford, who fought for lighter, inexpensive models, and Malcolmson, whose emphasis on heavier, more expensive vehicles aligned with existing national sales trends—and greater profit margins. Eventually, the disagreement proved irreparable, and the two split. Ford elevated the Model N, for the time being, into the workhorse of the Ford Motor Company.

Ford told the *Detroit Journal* that the $500 Model N had "solved the problem of cheap as well as simple automobile construction . . . within the reach of many." At the time, he considered the car "the crowning achievement of my life."[11]

In the fall of 1906, Ford hired one of his suppliers, Walter E. Flanders, to redesign the production processes for the Model N. Flanders reorganized the factory, assigning workers specific tasks and employing runners to keep the material moving. He established monthly production schedules, reorganized reporting structures, and through his efforts realized even greater profits through production efficiencies, but not a higher retail price. At a time when automobiles were assembled by hand, the Ford Motor Company sold nearly 8,500 cars, five times more than its best previous performance over any twelve months.[12]

A successful entrepreneur in his own right, Flanders left in 1908 to start his own company. The constant flow of talent was one of the costs of doing business in the early days of the automobile industry. With hundreds of automobile manufacturers suddenly competing for attention in the United States, many in the Detroit area, employees moved freely from one to the next in search of a small raise and better working conditions.

Ford was forty-five years old when he introduced the Model T in the fall of 1908, available in the early years in gray, blue, green, and red. Its impact on the American economy, culture, and infrastructure over the next fifteen years could not be overstated. The vehicle, and through it, Ford himself, emerged as a new industrial archetype of the twentieth century, standing up for those who lacked the means to stand up for themselves, those the Gilded Age had left behind.

Ford's obsession with affordability endeared him to the American people. Heralded for its mechanical features, the Model T Ford was an automobile for "poor people," a symbol of a new type of freedom, absent of pretentious luxuries and affordable by all. "There are a lot more poor people than wealthy people," he told reporters. "We'll just build one car for the poor people."[13]

Within days of its debut, thousands of orders rolled in, prompting relocation to a new factory in 1910, at Highland Park, Michigan. In three short years, the operation evolved into the fabled assembly line, the philosophical and operational foundation of the Ford empire for decades to come.

The automobile had come a long way in only a decade since Ford's Quadricycle puttered down the streets of Detroit. More reliable gasoline engines were replacing steam in factories, powering lathes, grinding and gear-cutting machines, and turning mills. Jobs were becoming repeatable, consistent, and scalable.

Innovative materials, engineering, and factory design combined to produce the Model T. The vehicle was built of light and durable vanadium steel; featured an in-line, four-cylinder engine that could achieve speeds of forty-five miles per hour; and could run on gasoline, kerosene, or ethanol. A rear-wheel drive, planetary transmission offered plenty of power on America's patchwork of dirt and gravel roads. Paved streets were still a rarity. A Model T Roundabout, the base model, was not cheap. It cost $825 in 1909, two-thirds the annual salary of an average factory worker. But it was less than most competitors charged and put the automobile in reach.[14]

The Ford Motor Company sold nearly 11,000 automobiles over the next year. Orders rolled in so fast the company stopped accepting them.[15]

❦ ❦ ❦

Ford eventually used some of his profits to purchase the Ford family homestead outside of Detroit. It became a source of inspiration, a quiet and familiar retreat from the relentless pounding, drilling, and hammering of the automobile factory in constant motion. The family farm also reminded him of the inefficiencies and "drudgery" that drove him away.

In the United States, the average turn-of-the-century farm was smaller than fifty acres, much too small for the impressive steam traction engines more familiar to farmers on the expansive prairies of the American West. Steam was more efficient, but dangerous. Boilers took hours to heat, building pressure that had to be regulated at all times to prevent violent and sometimes catastrophic explosions. Fireboxes were fed highly combustible material, typically wood or straw, on wide-open, wind-blown prairies.

Gasoline was less efficient and harder to obtain, but much more predictable than alternatives. Smaller, portable engines fueled by gasoline or less expensive

Early twentieth-century threshing scene in North Dakota.
Library of Congress, LC-USZ62-40414.

kerosene were becoming a more common power source for tasks on the farm, and a few manufacturers were offering gasoline traction engines, or tractors, to replace steam traction engines on large western farms.

Tractors were used for belt work or drawbar work. A pulley, connected to the flywheel, could be connected by a leather belt to another machine, transferring power from one to the other. The drawbar was a simple horizontal bar at the rear for attaching an implement, most commonly a plow. Tractors were hard to turn, required fuel that cost more than animal feed, and were built with hundreds of custom parts, which were themselves scarce. Parts, and the mechanics to install or repair them, took a long time to secure. And the most reliable, economical form of farm power was still in active service. No farm tractor yet could match the versatility or reliability of the horse.[16]

Ford, surrounded by small farms, had another idea. In 1906, two years before the debut of the Model T, he directed one of his engineers, Joseph Galamb, to assemble a rudimentary tractor built mostly from leftover automobile parts. It had four spoked wheels, was powered by a Model B automobile engine, and had a large, cylindrical water tank mounted on the frame, ahead of the steering wheel and between the two front wheels. A 1907 experiment took the form of what Ford called an "automobile plow," built with a 24-horsepower, four-cylinder engine and planetary transmission taken from a Ford Model B. The rear wheels came from a grain binder, with the front wheels, axle, steering, and radiator from the Model K.

Ford's concept was reinforced at a series of farm equipment demonstrations. To increase lagging attendance at a local fair in Winnipeg, Manitoba, Canada, in 1908, organizers decided to hold a tractor contest. It was met with an unexpectedly positive response. Over the next few years, trials evolved from a loosely organized demonstration into a must-attend event for top agricultural equipment manufacturers. For the first time, farmers could see both steam engines and gasoline tractors in action, the likes of which they had only read about. And they could talk to their designers and learn about potential applications on the farm. It was not the event's intent, but the demonstrations made it increasingly clear that the steam engine era was coming to a close.

Henry Ford sent the Farm Implement News *this photo of an early experimental tractor built from automobile and farm equipment parts, 1908.*

The annual demonstrations in Winnipeg further convinced Ford of the value of a small, gasoline tractor as a means to rid the farm of the drudgery and inefficiency that plagued him as a youth on the family farm. In Winnipeg, he saw huge, steam-belching engines mired in the mud, while their smaller, nimbler counterparts demonstrated easy maneuverability and greater efficiency.[17]

In October 1908, as the Model T began to emerge from the Ford factory, Henry Ford, a still relatively unknown local automobile manufacturer, submitted an article and photo to the *Farm Implement News*, a leading farm journal printed in Chicago. It was published on November 15.

FORD'S FARM TRACTOR.

> *One of the automobile concerns which is experimenting in the line of farm tractors is Ford Motor Company, of Detroit, Mich., whose president, Henry Ford, is said to be quite enthusiastic over the subject. The accompanying illustration is a reproduction of a photograph recently taken of Mr. Ford's farm. It shows one of the company's experimental farm motors drawing a disk harrow. The driver of the tractor is Mr. Ford.*[18]

For now, the only thing keeping Henry Ford from his farm tractor was the resounding success of the Model T.

2

JOHN DEERE: "THE NEWCOMER"

Eighteen-year-old Cyrus McCormick, Jr., left Princeton College in 1877 to join a team from his father's company, the McCormick Harvesting Machine Company, that took the company's latest product to Europe for field trials. An attachment called a wire binder had been developed for the harvester, a horse-drawn farm implement that cut, gathered, and ejected neatly collected bundles of wheat, oats, straw, or corn. The new binder device automatically tied the stalks, eliminating the work of two men who before would follow the harvester, gathering and manually tying the freshly cut crop into bundles, also called shocks or stooks.

Twine soon replaced binder wire, and the younger McCormick was given the opportunity by his seventy-three-year-old father to lead the transition—and protect it against competitors. His temporary leave from Princeton became permanent, and he returned to Chicago to work as his father's assistant until his father's death in 1884. Just twenty-five years old, Cyrus McCormick, Jr., was named president of the country's largest farm equipment manufacturer.

The young executive was a ruthless protector of profit, immediately pitting himself against his factory employees, lowering wages before his first year as president had even passed, despite strong sales. In a matter of months, all 1,400 employees were on strike. Union organizers were purged, and eventually McCormick was counseled by Chicago's meatpacking king, Phillip Armour, to settle the strike. The capitulation brought only a short reprieve. A new wave of strikes took place beginning in February 1886, culminating in the Haymarket riot, what many newspapers called the "McCormick riot," in early May.

The crowd of fewer than three thousand people had mostly dispersed when nearly two hundred officers arrived with their Winchester rifles. Rain had started to fall, causing some of the speakers to leave early. Then a bomb exploded in the crowd. Rifles were discharged in the smoke and confusion, killing at least six policemen and four workmen, the latter there to lobby for the eight-hour workday. Another thirty people were injured. Later that day, Mayor Carter Harrison called for Chicago residents to remain in their homes, with much of the nation going on lock-down the following day in anticipation of widespread riots that never came.[1]

Five organizers were eventually tried and hanged. Another killed himself or was murdered in his jail cell; the circumstances were never determined. Several others were eventually pardoned by the governor.[2]

Cyrus McCormick, Jr., was merely emboldened and now looked to eliminate competition in the harvesting business. Merger talks between McCormick and William Deering, his father's long-time competitor, and at least sixteen other harvester manufacturers fell apart in 1890 when the combined companies could not secure funding for the deal. Seven years later Deering, at seventy-one years of age, looked to retire, while his two sons, James and Charles, looked for a way out of the business. "With the great strides which [the Deerings] and ourselves have been making during the last two or three years," McCormick confided in his mother, "it is a fact that this concern and ours practically have two-thirds of the entire trade now, and large economies in operations could be made by working in harmony instead of by fighting each other as we are now doing."[3]

Talks ended three months later.

Over the next four years, after significant product expansions, plant investments, and an expensive marketing battle, the two embarked on a third round of talks, now with a clear path. Between 1898 and 1902, 212 consolidations had occurred in the United States, led by the formation of the $1 billion U.S. Steel Corporation. With a plan laid out, the companies called on Judge Elbert Gary, chairman of U.S. Steel, to represent the Deering interests, and George Perkins, partner of ruthless financier J.P. Morgan, to broker the deal.[4]

The result was the formation of the International Harvester Company in 1902, a merger of the McCormick Harvesting Machine Co., the Deering Harvester Co., and three smaller firms, the Plano Manufacturing Company, the Milwaukee Harvester Co., and Warder, Bushnell & Glessner. The new company consolidated 85 percent of the harvesting equipment market, easily the most profitable piece of equipment for both farmer and manufacturer.

The International Harvester Company was the fourth-largest corporation in America.

"The benefits that will undoubtedly result to farmers . . . have probably not been considered by a large portion of the farming community," an Illinois editor wrote, highlighting that only a combination would keep harvesting machinery affordable. "To maintain the present prices of these machines means to continue and increase the development of the agriculture of the world."[5]

The consolidated International Harvester reported net profits of $5.6 million in 1903 (legally they were two companies, the International Harvester Corporation and the International Harvester Company of New Jersey, split between domestic and foreign sales organizations), paying its owners dividends of $3.6 million. Sales grew to nearly $100 million by 1907, with products ranging from harvesters to manure spreaders, hay rakes and mowers, cultivators, pump jacks, stationary engines, and corn pickers. They were vertically integrated—owning and operating steel works, ore mines, and timber tracts, and manufacturing millions of dollars' worth of binder twine each year. Soon, the company was operating factories in the United States, Canada, Russia, Germany, France, and Sweden, and employing nearly 40,000 people worldwide.[6]

❧ ❧ ❧

A few manufacturers dominated the agricultural equipment business, which for nearly a century was divided between makers of harvesting equipment and of plows. That was about to change.

In the days leading up to Christmas 1909, bitter cold temperatures in Omaha, Kansas City, and Des Moines plummeted further. In some places, the thermometer read as low as 13 degrees below zero, with no end in sight. The city of Chicago was prepared for a mild December but instead faced bitter cold, a deep layer of snow arriving mid-month. Gas lines froze, plunging homes throughout the city into darkness. Influenza was rampant in both adults and children. Overheated stoves and furnaces created an unmanageable flow of calls to fire departments throughout the city.

International Harvester had just recently moved into its newly constructed, fifteen-story, neoclassical-style headquarters at 600 South Michigan Avenue, one part of a city-wide construction boom. The Harvester Building, neighbor to the freshly completed Blackstone Hotel, rented ground floor retail space at $1,000 per frontage foot. The rest of the building, of which Harvester operations took up only a part, was being "easily filled with a substantial and attractive list of tenants."

The exterior of the Harvester Building featured heavy stone ornamentation but was otherwise modest and unremarkable. Conversely, the interior offered modern technology and design, featuring highly polished surfaces, natural light, and high-speed elevators, all supported by a steel skeleton, literally the backbone that allowed Chicago's cityscape to reach to the sky. Its façade, plain and modest, was a proper representation of the company's rural clientele. The interior spoke to the efficiencies of the modern corporation, an emerging metaphor for the transformation of the twentieth-century farm.[7]

Less than two hundred miles west of Chicago, Moline, Illinois, which was "not as far from civilization as you may have thought it to be," a pleasantly surprised employee recruited to John Deere from Chicago wrote, was nestled on the banks of the Mississippi River, a unique spot where the mighty river ran east to west. For nearly seventy years it had been home to the world's largest manufacturer of steel plows, John Deere.[8]

Cyrus McCormick, Jr. (middle) and sons Gordon (left) and Cyrus (right), 1915. Wisconsin Historical Society, WHS-77376.

From his office in Chicago, Cyrus McCormick, Jr., picked up the cylindrical telephone receiver and pressed it near his thickly mustached upper lip, then turned the hand crank. His office was surprisingly plain: a roll-top desk pushed against the wall, an oak desk chair on wheels atop a small oriental rug, a small table, and a few tall ladder-back chairs for colleagues. He had just returned from Europe, visiting new farm equipment factories in Croix, France, and Neuss, Germany, and scouting locations for a new plant in Russia to build mowers and harvesters.[9]

A woman's voice came on the line and asked where to direct the call. "William Butterworth," he responded, a member of the Chicago Club, at the corner of Michigan Avenue and Van Buren Street. The stretch between the two buildings would soon to be transformed by the newly proposed "Plan of Chicago" that would, in time, turn Michigan Avenue into one of the most important commercial districts in the world.

William Butterworth, a man of "broad, penetrating vision," was a frequent visitor to the city. Unlike the long-tenured McCormick, Butterworth was only two years into his time as the president of Deere & Company, the long-time plow manufacturer better known by its trade name, the proper name of its founder, John Deere.[10]

Both companies had just reported record years. International Harvester's $15 million in *profits* tripled Deere's total sales of $5.2 million. In fact, Harvester paid out nearly as much in stock dividends as Deere had earned in total sales.[11]

Outsiders had been underwhelmed by Butterworth's ascension to the head position at John Deere after the death of his universally respected father-in-law, Charles Deere. "It appears that the organization is somewhat unfortunate in that the present president of Deere & Company, William Butterworth, is new to the business," wrote a representative from the United States Bureau of Corporations, the organizational predecessor to the Federal Trade Commission, noting that "competent outside authority considers that the Deere organization is much weakened by Mr. Butterworth's lack of experience.[12]

Those around him knew otherwise. William Butterworth was raised on a farm in Maineville, Ohio, northeast of Cincinnati, but spent considerable time in Washington, DC. His father, Benjamin, whose ancestors came to Pennsylvania with William Penn, rose from Assistant United States District Attorney to state senator of Ohio, then to United States congressman. Benjamin twice served as the United States Commissioner of Patents, from 1883 to 1885, then again from 1897 to 1898.

While attending school, William worked as secretary for the U.S. Commissioner of Patents, W.E. Simonds. Upon passing the bar exam, he married Katherine Deere, the older sister of his best friend's wife, in 1892. Her parents built them a house across the street from their own in Moline and offered Butterworth a job.

Charles Deere, a "man of pluck and purpose," trained his son-in-law for the next fifteen years, teaching him to balance personal and sometimes conflicting business interests. As a result, Butterworth was financially cautious with the family money that still financed the company, pushing for long-term gains in a cyclical, low-margin, weather-dependent business. Personally, he was exceptional at everything he ever tried. At Lehigh University he excelled in sports and the fine arts, playing middle linebacker on the football team, pitching for the baseball team, and performing with both the Glee Club and the Dramatic Club. Parties at the Butterworth home were filled with music, dancing, and laughter.

A company engineer recognized the pressures of a family business, noting that Butterworth "understood [Deere] well and constantly, [and] whenever possible, Charlie Deere pumped information into him of various kinds and in particular as to how the business of Deere & Company should be managed, . . . And Will Butterworth usually suppressed his own notions and preferred to listen to the ideas of his associates, always comparing them with the Charlie Deere motives, to which Mr. Butterworth was always loyal."

Outsiders mistook Butterworth's preference for steady forward progress as indecision, failing also to understand the cyclical and unpredictable nature of the agricultural equipment business. Manufacturers relied on favorable weather and strong commodity prices as a counter to the volatile prices of steel, wood, and other raw materials, the purchase of which was done a year in advance, based solely on next-year sales forecasts. Each year was its own series of complex, calculated risks, each dependent on the other.

Unforeseen by most, Butterworth was up to the challenge.

Farm tractors were not on the agenda for Cyrus McCormick and William Butterworth as the two discussed competing lines. Harvester had only started tractor development, and Deere, which had formed selling arrangements with a few manufacturers, had not considered production. There were only nine registered tractor manufacturers in the entire United States, with total production of only two thousand machines.

Instead, McCormick's call to Butterworth was a call to action by Alexander Legge, Harvester's "kindly," yet "brusque" assistant general manager. His friends called him the "iron man" for his ability to work endless hours whether in the bitter cold or extreme heat. Legge had learned that

Butterworth had recruited a Harvester employee, which the Deere president acknowledged and then defended as a consequence of the natural state of business. Legge, through McCormick, considered it a competitive escalation. Deere, Legge confirmed through his own sources, was implementing plans to enter the harvesting business and for the first time planned to compete directly with International Harvester.[13]

Legge's information was accurate. Butterworth and the team from Deere were fresh from failed merger negotiations with the J.I. Case Threshing Machine Company, one of International Harvester's remaining competitors, and then "terminated unfavorably" a potential acquisition of the Frost & Wood Co. of Smith Falls, Ontario, Canada."[14]

McCormick attempted to assuage Butterworth not "as the President of our Company but as an acquaintance," hoping to ensure that the two did not "get their wires crossed . . ."[15]

Through a plan formulated by Legge, Harvester would distribute John Deere plows in Canada in exchange for Deere's commitment to stay out of the Canadian binder business. "For some years our traveling men and our agents had been pressing us very strongly to go into the plow business," McCormick told Butterworth, "but we had up to the present time strongly declined such a policy." Instead, Harvester took on lines where they did not "have to invade the territory of our friends and allies in the implement business."[16]

The two companies had avoided direct competition for more than seventy years, but "of course if the plow people went into the harvester business, we might find ourselves drawn into the plow business," McCormick warned.

The phone meeting ended without resolution, and the two met for lunch soon after, neither fully revealing his future plans, nor all that had been accomplished privately up to that point. Harvester leadership discussed buying Deere out, perhaps forming a new subsidiary and building plows on their own. They could form a partnership, or even make an "agreement as to each keeping out of the other's business (make a truce for three or five years)," relayed an internal memorandum. In anticipation of Deere balking, Harvester had already negotiated exclusive distribution rights for plows built by Canton, Illinois, manufacturer Parlin & Orendorff, a long-time Deere competitor.

Butterworth already knew about the arrangement prior to their phone call.[17]

Privately, Butterworth actually agreed with McCormick on most points. "It would be the height of unwisdom to attack the International Harvester Company," he told his own team. "A serious fight with them will mean a serious depletion of our profits, maybe their absolute curtailment, thus endangering our dividends, the passing of one of which would very seriously affect its market price and do great damage to our credit."[18]

Yet, the meetings with McCormick only bolstered Butterworth's resolve. On his return to Moline, Deere's board of directors was urged to take "the matter up vigorously." To protect Deere's Canadian trade, they would expand into "binders, mowers, and kindred lines of goods . . ."[19]

Aggressive and sweeping reorganization plans at John Deere were consummated in less than a month. A year later the work was nearly accomplished. Willard Velie, John Deere's grandson, long-time board member, and president of the Moline-based Velie Motor Corporation, brought the board a number of successful negotiations ready for approval. Their work started with the Moline

Willard Velie (left), Dr. Warren E. Taylor (middle), and William Butterworth (right). Courtesy John Deere.

Wagon Company, makers of wood and steel frame wagons, which had "indicated a willingness to dispose of their entire property."

Considerations for acquiring other long-time partners were accelerated, as was the renewal of contracts with independent implement dealers throughout the country and a further "amalgamation of all the interests allied with this Company . . ." Deere & Company had interests in branch houses, independent feed stores, and implement companies, and even in competing product lines. The time had at last come to rebuild John Deere based on the prevalent organizational models of Standard Oil, U.S. Steel, and proven attainable by the International Harvester Company. John Deere would make and sell everything needed on the farm. The industry called it a "full line."

On January 6, 1910, the board of directors appointed five men to follow through on a resolution passed earlier that meeting. "Whereas, the directors and stockholders of Deere & Company view with much concern the aggressions of competitors and feel that for the conservation of their investments and interests a unification of their allied factories and distributing houses should be perfected."[20]

The plan was straightforward. First, consolidate existing sales branches and co-partnerships, then acquire complementary product lines, and last, add additional lines. The work was now to be formally led by a group internally known as the Committee of Five. Their first order of business was to fully acquire the Deere & Mansur Company, a forty-year-old co-partnership that built corn planters and haying equipment in Moline. Consolidation of existing sales branches followed. Three days of meetings in the final week of April 1910 finalized the next phase, a more ambitious plan. That summer, Butterworth told the board that the acquisition of the Kemp & Burpee Manufacturing Company of Syracuse, New York, and execution of a plan to build a factory in Welland to manufacture grain binders for the Canadian market were "as incidents of necessary expansion." He nominated George Mixter for the new position of vice president in charge of general operations, with responsibility for the "allied factories" of Deere & Company: Union Malleable Iron Company, Marseilles Manufacturing Company, Kemp & Burpee Manufacturing Company, and the projected Canadian binder factory.[21]

Young and ambitious, Mixter began the work of integrating new acquisitions, rationalizing production, and building new, modern factories. "The problem of bringing these plants into harmonious relation with Deere & Company was not easy," he later wrote, "and practically not entirely accomplished by the time I ceased my relations with Deere & Company in 1917."[22]

The most complicated issue remained the grain binder, and through it, an unwinnable battle with International Harvester. Fortunately, an existing ally, the Dain Manufacturing Company of Ottumwa, Iowa, offered a solution. Its operations included an existing manufacturing facility in Welland, Ontario, Canada, a point of entry for Deere's plans.

Joseph Dain was a well-dressed, handsome entrepreneur, "naturally of an inventive nature" and of "pleasing personality," whose mechanical mind and equal ease in the board room put him always on the precipice of the next big thing. He was also highly regarded by his employees. He was born near London, Ontario, Canada, moving to northern Missouri as a child. Eventually, he entered the furniture business in Meadville, but after years of watching area farmers struggle with what he considered to be slow, outdated methods of managing their hay crops, he set out to change the process.

Dain's first invention was an automatic hay stacker, followed by a new type of sweep rake. Reminiscent of McCormick's earliest demonstrations of the reaper, farmers initially ignored the new tools. Dain bet on himself, selling his furniture business and starting production in a small shop in Springfield, Missouri. In 1882 he received his first patent, and in 1887 established the Dain Mower Company in Carrolton, Missouri. Three years later he incorporated the Dain Manufacturing Company for $40,000. By 1900, capitalization had been increased to $600,000, and the Dain line included hay stackers, sweep rakes, and mowers, soon after distributed by John Deere on a non-exclusive basis.[23]

When Deere's Re-Organization Committee met on November 28, 1910, they read a first draft of the proposed acquisition and consolidation of six companies, many of which had already been executed, with a few more lines, manure spreaders, and wagons, to add. Another, the Syracuse Chilled Plow Company in New York, approached Deere about acquisition. Its owners were asked to wait, which they did, and Deere eventually purchased them as well.[24]

❦ ❦ ❦

The *Farm Implement News* publicly broke the news of Deere's ambitions. The illustrated cover of the August 4, 1910, edition featured a lean, athletic deer, antlers forward, in mid-leap through an open gate. To its left stood an enormous elephant, the letters I.H.C. on its side. To both right and left, gazing at the deer, were two horses and two steers with the names of smaller harvesting companies, Acme, Johnston, Wood, and A.P. Small print underneath the illustration announced "The Newcomer in the Harvest Field."

Despite the wave of acquisitions and consolidations, Butterworth was not fully committed to an inevitable, all-out competitive war with International Harvester, which had seven dollars to every one of Deere's in working capital. A more aggressive entry into the harvesting business required capital Deere did not have in the surplus. That meant borrowing, which possibly meant outsiders on the board. Protecting the interests of the family-controlled business, Butterworth thought it nonsensical to act on the "theory that because the Harvester Company may go into the plow business, we should go into the harvester business and so start a fight with them." But the landscape continued to evolve rapidly.

Deere executives were contacted by government agents in a pending investigation of monopolistic activities by the International Harvester Company. Before the year ended, both the Standard Oil Co. and the American Tobacco Company, two of the nation's largest companies, would be forcibly broken into smaller pieces under the recently passed Sherman Anti-Trust Act, which sought to protect competition in American industry. After much internal debate, Deere responded to the inquiries from the Department of Justice, offering that International Harvester's "practices are unfair in a number of instances, but extremely fair in others."

Internally, the potential dissolution of the International Harvester Company put Deere's plans into a new context. Despite Butterworth's initial reservations about American binder production and sales, direct competition with Harvester now looked like the necessary capstone of a larger plan to reorganize and modernize Deere & Company.

Willard Velie, among others, was confident that International Harvester's breakup would come to pass. "There is no such thing as equilibrium in business," he wrote, and they had to choose to be "progressive or retrogressive."[25]

Velie won the day. The profit of the consolidated Deere companies came to nearly $4 million at the end of 1911, followed by a modest increase the next year. A $10 million preferred stock offering infused more capital into the business. With a motion to invest some of the proceeds, and finding that their experimental binder design had been "carefully weighed and found *not* wanting," material for the assembly of 4,000 binders was approved—in a new factory in the United States.

Deere hired A.C. Funk, general superintendent of International Harvester's Champion Works, in Springfield, Ohio, and Harry Podlesak, a designer at Harvester's McCormick Works in Chicago. Funk left shortly after for health reasons, and Deere lured Willard Morgan, a former sales manager at International Harvester with nearly thirty years' field experience, to join them.

Harvester's McCormick Works and Deering Works, both in Chicago, had combined capacity for 675,000 binders, reapers, mowers, rakes, and drills. At Deere, assembly of a newly designed grain binder got underway in the neighboring town of East Moline in rented buildings of the recently, and perhaps ominously, bankrupt Acme Harvesting Company. While the buildings were readied, assembly began outside under the cover of rented canvas circus tents. Five hundred machines were ready for the fall harvest.[26]

Though much work was to yet to be done, Deere finally had the makings of what it considered to be a full line of agricultural equipment. Ground was broken on a new factory for the production of harvesting equipment for corn and grain in East Moline, just a few miles upriver from the John Deere Plow Works in Moline. Within two years, production exceeded four thousand units.

At last, two companies at the leading edge of the nineteenth century's agricultural revolution, increasing yields through technology and constant product improvement, were in direct competition in nearly all product categories. But at Deere, one piece was still missing. There was also little indication they planned to do anything about it.

The Harvester World, International Harvester's monthly company news magazine, featured a comic on page six of the July 1910 edition.

Charitable Person—"I thought you were blind."

Beggar—"Well, Cap, times is so hard just now and competition is so keen that even a blind man has to keep his eyes open nowadays if he wants to do anything at all."[27]

Both William Butterworth and Cyrus McCormick, Jr., knew that harvesting equipment was only the beginning. In fact, the "full line," the phrase used to reference a company's ability to manufacture every piece of equipment used on the farm, was about to be redefined. Once again, with modest investments into tractor development underway, International Harvester had a head start.

A month before breaking ground on the John Deere Harvester Works in March 1912, John Deere's board of directors passed a resolution, committing themselves to the development of a new machine. They called it a "tractor plow."

3

THE TRACTOR WORKS

Automobile manufacturers, truck makers, and farm equipment manufacturers wrestled with similar challenges as they worked to manipulate the power of the internal combustion engine. There was a surplus of possibilities but few success stories. Experimental machines were hidden in garages, sheds, and in plain sight on city streets, but rarely ever went further than the erection of one machine. Bankruptcy was the most common outcome.

Mechanical parts were often crude and wore down under constant operation or broke down completely as parts rubbed together or worked their way out of alignment. Starting cranks were unwieldy and often kicked back, causing trauma to limbs. Engines misfired and belts slipped. Roads and farm fields were uneven, rocky, muddy, or dusty, causing flat tires and axle and chain breakages. Fuel had to be purchased well in advance, as there was no infrastructure in place for refilling on the go. In all cases, the operator served as mechanic.

On the surface, the design, function, and engineering principles of automobiles and tractors were similar. The end product was based on the premise that the internal combustion engine could be adapted for jobs performed by horses or steam power, whether that be pulling, pushing, lifting, or simply driving something forward and backward. Every day gasoline engines, large and small, were replacing their steam-powered ancestors in factories and farms, running everything from heavy industrial equipment to washing machines, irrigation pumps, and electrical generators.

Carriage and farm wagon manufacturers, which included International Harvester and John Deere, were part of the early wave of automobile innovators. Harvester's first automotive engine, not uncommonly, pre-dated its earliest tractor development. The engine was developed by E.A. Johnston, a machinist at the pre-consolidated McCormick Works, in 1897, less than a year after Henry Ford first tested his Quadricycle on the streets of Detroit. Johnston designed a second two-cylinder version the following year, installed it into a chassis, and for several months drove the five or six miles between

Edward. A. Johnston's auto mower placed first at the Paris Exposition in 1900. Wisconsin Historical Society, WHS-39487.

his home on Millard Avenue and the McCormick Works at Blue Island and Western Avenue, on the south side of Chicago. By 1905, the engine evolved into what he called an "auto-buggy," which Johnston designed and built at Harvester's Keystone Works, one hundred and twenty miles west of Chicago in Sterling, Illinois.

That year Johnston was further tasked with the creation of a farm tractor, though the definition of such a machine was still up for debate. The expiration of patents on the four-stroke-engine design of German engineer Nicolaus Otto created a race for production, with nearly four hundred different manufacturers now building internal combustion engines in the United States alone, many based on Otto's design. Focused on powering massive factory machinery, and for some small, on-farm applications, few companies had yet attempted to develop the engine for mobile power on the farm. Those that had made the attempt followed familiar steam engine designs, crudely constructing machines from a maze of parts designed for other applications.

Johnston's initial work, built from the working knowledge of his auto buggy, also took place at the company's Keystone Works. As none of the shops were yet outfitted for significant engine, tractor, or automobile production, Johnston soon coordinated work at three additional factories in Akron, Ohio, Milwaukee, Wisconsin, and Chicago. Early experimental machines were built from purchased parts, customized, and shipped to Sterling for further modification and assembly.

In 1906 and 1907, Johnston secured a single-cylinder stationary engine from the Ohio Manufacturing Company in Sandusky and mounted it on a truck frame. It developed into what they called the Type A tractor, which they built in twelve-, fifteen-, and twenty-horsepower versions. The Type A, which initially had its power transmitted from the engine to the drive wheels through friction and gear wheels, was replaced by a more durable gear-and-chain transmission, also developed by the Ohio Manufacturing Company. They called it, logically, the Type B. Simultaneously, a Type C model was developed with a single-cylinder engine and full-gear transmission.

Johnston began to assemble a team of engineers and mechanics who worked around existing production schedules. From the McCormick Works

A twelve-horsepower International Type A tractor with an "Ohio" hay press, 1910. Wisconsin Historical Society, WHS-70814.

in Chicago, H.B. Morrow oversaw tractor assembly in Akron, which "were made in considerable numbers." By the end of 1906, that amounted to fifteen tractors, earning International Harvester a leading position in the field, though well behind the Hart-Parr Company of Charles City, Iowa, whose thirty-horsepower No. 1 debuted in 1902. In 1906, the company built nearly 150 tractors, more than half of those available nationwide, in a variety of sizes. The company had also first popularly adopted the term *tractor* as a short-hand term for "traction engine" or the commonly used "agricultural motors."[1]

Orchestrating personnel, materials, and temporarily idle machines, separate projects were consolidated at the McCormick Works in February 1907, and then later that year transferred to the Akron Works. The combined effort led to a new 45-horsepower tractor by the end of the year. It would come to be called the Mogul, likely named after a popular type of powerful and well-known steam locomotive. Over the next six years, no fewer than thirteen different models were developed with one-, two-, three-, and four-cylinder engines offering as much as sixty horsepower.[2]

The Winnipeg Industrial Exposition had been held annually on an eighty-acre plot in North Winnipeg, just west of the Canadian Pacific Railway Yards, since 1891. Annual attendance had grown to more than 200,000, even before the 1908 tractor trials in what organizers billed as the Light Industrial Agricultural Motor Contests. Only two years after construction of its first machine, the Model A, International Harvester entered the 1908 trial. In time the event would be recognized as a watershed moment in the evolution of the farm tractor. Two years later the show would attract Henry Ford.

The International Harvester Company, the Transit Thresher Company from Minneapolis, Iowa's Hart-Parr, and Kinnard-Haines hailed from the United States, while entries from Marshall & Sons of Gainsborough, England, and Universal Motor Syndicated of Saskatchewan, Canada, represented the rest of the world. Models ranged in design from one-, two-, three-, and four-cylinder models and were given points in nine categories by two judges. Categories included turning and handling ease, hauling capacity (which was tested by plowing), length of operation without refilling the gas tank or adding water to the radiator, and price.

Rain hampered the tractor contests from the start. "The rain, which handicapped the engines so greatly in getting to the field, helped them fully as much when they started to plow, for the sod turned over like cheese and . . . the engines were able to travel right along and do very satisfactory work even if the field did contain considerable Red River gumbo."

The twenty-horsepower entry from the Transit Thresher Company broke down early and never competed. Organizers added a seven-ton weight limit in an effort to separate the competing gasoline tractors from their steam-powered counterparts, disqualifying the Hart-Parr, which was now too heavy. Its salesmen held a private demonstration across the street.

After three days of work in the scattered rain and mud, the Kinnard-Haines "Flour City" tractor took home first prize, followed by the entries from Harvester, a single-cylinder, ten-horsepower tractor "of their usual type of fixed or portable gasoline engine mounted on a truck" and a forty-horsepower tractor of "new design, having three cylinders placed horizontally . . ."

According to another local report, "The winner wore an air of 'what else could you expect,' while the others took their medicine like men should do, and congratulated one another."[3]

Fresh from a successful trial in Winnipeg, Harvester began work on a new two-cylinder tractor at its Milwaukee Works. The design incorporated an automobile steering mechanism and was ready for demonstration at the 1909 Winnipeg contest. Designers had just enough time to test two tractors before loading them in railroad cars for Winnipeg, painting them on the train during transport. After winning first-place awards in two classes, the tractors were further modified. By late 1910, in a single-story building at the McCormick Works factory on Twenty-Sixth Street in Chicago, the "Type D," soon after known by its trade name Titan, was ready for production.[4]

Meanwhile, developmental work on a separate design, named the Mogul, continued in Akron, while in the former Milwaukee Harvester Company buildings in Milwaukee, now with foundries and production of small stationary engines, cream separators, and more, engineers Leonard Sperry and Charles I. Longenecker worked on an entirely separate line of Titan-branded tractors in the ten- to sixty-horsepower range.

The twenty-horsepower Mogul Model C appeared commercially in 1909, catapulting Harvester to industry dominance with sales of more than 2,400 tractors over the next five years.

The original McCormick Works at Blue Island Avenue, Western Avenue, and 31st Street in Chicago, was, like most of the city, burned to the ground during the Great Chicago Fire of 1871. Its replacement rose like a phoenix from the ashes as a model of American factory design. By 1910, it was a city unto itself.

The main building was repurposed, but a dedicated network of red brick buildings and railroad tracks went up in a careful series of phases through 1916. But when its first occupants arrived, there was little to see. On January 30, 1910, Johnston, his assistant, H.B. Morrow, and a team of hand-picked designers all arrived. The shop was already outfitted with a thirty-five-horsepower motor, installed with line shaft, eight lathes, and an old shaper. Most of the initial equipment was "obtained out of scrap machine pile at [the] McCormick Works." A small drafting room and office were set up on the west side of the building. Two roll-top desks, a typewriter desk, and a No. 10 Remington typewriter arrived on February 2. Next, they cut a hole in the Blue Island Avenue side of the building—large enough to fit a tractor through.

"With Mr. E. A. Johnston as Superintendent and Mr. Morrow as his Assistant, it has always been a mystery to me how, during the first two years any of the foremen, especially the assistant superintendent, survived," noted one of the superintendents who did survive the next two years. "There was the forming and breaking in of an organization, buying of machinery, setting it in place, hiring men, keeping up production at all times, getting castings without pattern equipment, and many other heart-breaking trials."

On February 10, 1910, they adopted a new name. They called it simply the "Tractor Works." *The Iron Age*, a monthly publication tracking American industry, reported that $1 million had already been invested in the factory.[5]

Over the course of the next year, new buildings of the "most progressive types" went up, including a machine shop, forge shop, paint shed, and warehouses. The machine shop rose to sixty feet in the center, thirty feet high on the sides, with windows running the full length of the north side of the roof in "sawtooth roof construction" fashion. A twelve-ton crane spanned the bays on each side. Light work was done on the second floor, with heavy work supported by the five-foot deep cement floors below. The forge shop was of a similar design.

Two boilers were installed to help heat the buildings, supplementing the exhaust steam from the McCormick Works, which had proven inadequate to heat the plant. A test track was built on the north end, graded with limestone, and in itself became a curiosity to all those who saw it.[6]

Railroad spurs scheduled for the following year would soon take train cars for loading directly to the assembly building.

As spring arrived, they were making a tractor a week, the heavy sounds of cutting, punch, and planing machines vibrating throughout the buildings. Soon they could assemble a full tractor a day. At capacity, they could build three forty-five-horsepower tractors daily, and by the time construction and outfitting was compete, orders called for six a day. A night crew was added, and by 1912, after additions to the forge shop and warehouse buildings, crews were building fourteen tractors every day.

Johnston continued to show that he was a brilliant mechanic, but the frantic pace of development demonstrated that he was an even better manager of the work. "We made many errors and he told us how to correct them," an early superintendent noted. "We committed many acts of foolishness and often received a good calling down, and when we were discouraged, he encouraged us always."[7]

Harvester's full line showed no signs of slowing, with more than half of the 4,000 gasoline tractors sold in the United States in 1910 rolling out from their factories. Nearly 3,000 tractors, in a variety of horsepower ranges, were sold and shipped in 1911, as total industry sales reached 7,000 for the first time. Five new companies entered the field, followed by eleven more in 1912.

International Harvester was always prepared to defend its territory and prepared itself for new competition in a rapidly evolving market. "Fifty important firms on this continent are building tractors," wrote L.W. Ellis and Edward Rumely in their 1911 book *Power and the Plow*. "A million-dollar addition in Indiana, a new million-dollar plant in Chicago, a two-million-dollar factory in Iowa, have been erected to construct gas tractors in the brief interval since they have been recognized as a possibility. New companies are appearing, old firms expanding, to take care of the business that rewards aggressive methods."

"It takes four or five years to make a good work horse," wrote Ellis and Rumely. "A modern factory can turn out a 30-horsepower tractor in three to five hours. It takes many generations to change the types of an animal, but only a few weeks to adapt a machine to a new condition . . . the dominant form of power in our dry farms of the future will be gas power."[8]

The authors failed to take into account that none of those modern factories existed, though such a future was closer than even they knew as a wave of opportunistic entrepreneurs, existing farm equipment manufacturers, and savvy investors looked to capitalize on the promising new market. Agriculture, as it had for generations, was the foundation of the nation's economic prosperity, and a growing nation, fueled by the increasing embrace of emerging technologies in homes, factories, and farms, could only benefit from a new era of power farming. It appeared that International Harvester, not surprisingly, was positioned to deliver the machines to make it all possible.

4

"DIVORCE THE PLOW FROM THE TRACTOR"

A twenty-five-horsepower version of Harvester's Mogul tractor, the power equivalent of a team of four to six horses, was introduced in 1911, reaching annual production of nearly one thousand units. Farmers knew what a horse could do, but the farm tractor was complex, confusing, and unreliable. Early tractor models most commonly took the name of their company, such as the Huber or the Hart-Parr, supplemented by two numbers to denote horsepower. The Heider, for example, was a 15-27 tractor, meaning that it achieved fifteen horsepower pulling a plow from its drawbar, an attachment point commonly at the rear of a tractor, and twenty-seven horsepower when a belt was attached to the pulley to provide power for another machine.

Over a ten-year period through 1918, at least ten different Mogul models, "the tractor that left the disappointments out" according to a 1915 advertisement, were offered by International Harvester—a Mogul 30-60, Mogul 15-30, Mogul 10-20, Mogul 8-16, and a bevy of one- and two-cylinder experimental versions. Expectedly, Harvester had quickly emerged as the industry leader, delivering 40 percent of the farm tractors purchased by American farmers in 1911.[1]

William Butterworth and his colleagues at John Deere recognized the need to develop a farm tractor but understood, perhaps too well, the risk involved in bringing new technology to market. To date, the sizable investment, fragmented state of the market, and unique needs of farms depending on geography, soil and crop type, size, and more, posed more risk than return. They only had to look around Moline and to the growing list of expensive failures of budding automobile entrepreneurs. Even Charles Deere, Butterworth's father-in-law, had been victimized.

Possibly the first automobile in Moline was built in 1899 by Samuel Arnold in his garage. Powered by a dry-cell battery of his own design, Arnold's electric automobile was met by "surprise . . . wherever it went," according to *The Weekly Mail.* While Henry Ford was forming the Ford Motor Company in 1903, William E. Clark was already on his third venture in East Moline, now selling two versions of the single-cylinder car he called the Blackhawk. It too failed. But in time, Clark's work attracted an investment from long-time Deere president Charles Deere. The newly formed Deere-Clark Motor Company bought the entire factory of a bankrupt car maker in Lansing, Michigan, shipped it to East Moline, and began turning out two models, the Type A and Type B. Rapid progress ended abruptly with Deere's death in late 1907. Another Deere executive, Charles Pope, reincorporated the business into the Midland Motor Company. Time would tell whether it would survive.

After Deere's bold entry into the harvesting equipment business with a line of mowers and grain and corn binders, Butterworth continued to focus his energies on the protection of Deere's new investments for the farm. Others, led by the chairman of the executive board, Willard Velie, began to push for development of a John Deere tractor.

Velie was John Deere's grandson and joined the company after the death of his father in 1895, taking over his position as Deere's corporate secretary. Separately, Velie began building carriages in 1902, leveraging Deere's long network of branch and sales offices for distribution. In 1907, with his older brothers Steven and Charles working in the field at Deere sales branches, Willard was named vice president at Deere & Company, then soon after named chairman of a new executive council investigating organizational structure and decision-making authority for the company as it expanded. Velie was often at odds with his cousin by marriage, William Butterworth, and was a strong voice for decentralization.

Velie's conversion from horse to engine power was transformative, as he recognized early that the days of the horse-drawn buggy and carriage were drawing to a close. The Velie Motor Vehicle Co. was formed in the summer of 1908. His automobile debuted at the Boston Automobile Dealers Association show in April 1909, six months after the debut of the Model T.

Velie followed the path of the many auto manufacturers that first tested their designs on the racetrack. Six months before its showroom debut, in August 1908 a Velie participated in the first competition ever held at the Indianapolis Motor Speedway. Despite a "noble, and in fact, heroic effort," his driver, J.H. Stickley, had to be extracted from the number forty-nine car and excluded from the fifty-mile race. Heat exhaustion after the previous day's events "was too much for him." Velie returned year after year, earning a seventeenth-place finish in 1911.[2]

Velie's lack of focus on farm equipment caused dissension among Deere's board, especially from sales manager George Peek, his cousin. Still, Velie pressed his opinion that the internal combustion engine held an inevitable and permanent place on the farm. His own growing automobile sales, now reaching more than three thousand vehicles annually, provided ample evidence of technology adoption. That summer he told the board of directors' executive committee they "had better consider the matter of securing selling alliances with manufacturers of tractors."

Velie, a carriage, and now, automobile manufacturer, had spent most of his life around farms and farm implements and knew first-hand the very different challenges posed by the tractor. To bolster his case, he found an ally in Dr. Warren Taylor, the newly hired manager of Deere's many experimental farms in East Moline and Moline. Taylor was equally as progressive in his study of agronomy, what was then called "soil culture," and was positioned to give new technology a real place in the soil.

On March 5, 1912, Velie successfully moved the tractor question forward, and with mixed support, the executive committee passed a resolution to build a "tractor plow." By that time, the full company reorganization was nearly complete.

> *RESOLVED, That in view of the inevitable future use by farmers for diverse purposes of gasoline and kerosene tractors, and especially since the trend is to use them in connection with implements, particularly plows, it seems vital to the interests of the Company that serious cognizance should be taken of the situation, and that through its experimental department, the personnel and talent of which shall be increased, if necessary, a movement to produce a tractor plow should be started at once, having in view constantly, that the success of the same would be enhanced if not assured, were it possible to divorce the tractor from the plow and to thus make it available for general purpose.*[3]

Plowing was the foundation of a healthy and profitable crop, the first task of each growing season, executed to create a healthy environment for plants to grow. Deere had been advancing the process since 1837 when John Deere himself built a self-scouring steel plow, designed to conquer the dense, clingy Illinois soil.

Plowing was accomplished at different times of the year depending on geography and crop type but in all cases was a basic process of pulverizing the soil and turning over trash, which was a reference to leftover crop residue (for example, the bottom part of a corn stalk still embedded in the ground) to create a seed-bed. Plowing aerated the soil, creating pockets for air and moisture absorption.

Plowing was both subtle and complex, each plot of land unique, even for adjoining properties. Clay soil could not be plowed when wet or it would clump. Sandy soil required an entirely different technique. Virgin soil, as was still found out west, required yet a different approach for the first plowing, and that was then revised the following year. The breaking of virgin soil created the early market for lumbering steam engines and early gasoline tractors powerful enough to pull ten or twelve gang plows. But after the first year, the tractor had diminishing value.

The walking plow, an evolution of an ancient instrument, still required the operator to walk behind the horse, guiding the plow through the ground with two wooden handles. It was still the single most predominant implement on most farms. Over the years, riding versions, which put the operator on a seat, had become more commonplace. Early versions were in part designed for Civil War veterans without use of their arms and legs who were returning to the farm. Even into the early twentieth century, plows were still designed for horse-farming.

Tractor plows, as Deere had learned through several collaborations with tractor manufacturers, required yet another evolution of design.

Deere did not dismiss the tractor but was methodical in its investigation, first working to determine whether to buy an existing manufacturer or begin internal development. In fact, Deere had been connected to the farm tractor and its predecessor, the steam traction engine, for half a century. In 1858, John Deere himself watched a thirty-horsepower steam plow, one of the first successful steam plows in the United States, in action at the Illinois State Fair. There Joseph Fawkes's American Steam Plow pulled eight, fourteen-inch prairie plows, turning furrows at a blistering—for a tractor—four miles per hour. The frame on its own was eight feet wide and twelve feet long. On his return home, Deere set to the work of building one himself, telling the newspaper that "it will be a great day when Illinois can show a steam engine taking along a breaking plow, turning over a furrow ten or twelve feet in width as it goes."

The next year, Fawkes's steam plow won a gold medal from the United States Society at the Chicago Fair. Presidential candidate Abraham Lincoln saw the benefits of steam power for farm work, foreshadowing the struggles of gasoline tractor manufacturers fifty years later. He told a crowd at the Wisconsin State Agricultural Society that it was not enough "that a machine operated by steam will really plow. To be successful, it must, all things considered, plow better than can be done with animal power." It had to do the work just as well, he said, cheaper, faster, and without failure.

Deere's efforts to build a steam plow were quietly abandoned.[4]

In 1879, Deere, Mansur & Company, Deere's Kansas City sales branch, sold the Watertown steam engine, and in the early 1890s, another branch sold Star Traction Engines from C. Aultman & Co. of Canton, Ohio. Partnerships continued, but in-house manufacturing had always been considered an expensive, and unprofitable, gamble.[5]

Deere now took that perspective into new investigations, beginning with a visit to the Gas Traction Company in Minneapolis. The two companies had partnered at the Winnipeg trials in 1909, the Gas Traction Company's Big Four 30 tractor and a fourteen-bottom John Deere plow winning the gold medal for the over-thirty-horsepower class. The Big Four 30 was a giant among tractors of the day, weighing in at 19,000 pounds, with drive wheels reaching eight feet in diameter. The four-cylinder engine was a self-proclaimed industry first, the work of a thin, spectacled Minneapolis pastor-turned-farmer named Daniel Hartsough. He was of a new class of independent entrepreneurs looking to disrupt the agricultural machinery industry.

North and South Dakota were very much the proving grounds for the first generation of tractor manufacturers rich in their knowledge of steam-powered machines. The Homestead Act, passed in 1862, encouraged western settlement. For $1.25 an acre, settlers could claim 160 acres as long as they built a residence, cultivated at least ten acres, and remained for five years. The Timber

A Big Four 30 tractor sits idle with three grain drills. Courtesy John Deere.

Culture Act of 1873 provided additional incentives, and an expanded Homestead Act increased claims to 360 acres. The continued annexation of tribal lands by the United States government displaced the native tribes that had cultivated the land for generations and now brought hundreds of thousands of additional homesteaders who could acquire the land by lottery. Combined, the populations of North and South Dakota reached 750,000 people by the turn of the century. By then, many of the original homesteaders had fulfilled their five-year requirements and sold their property, giving way to massive farming and ranching operations perfectly suited for steam traction engines and powerful gasoline tractors.

Restlessly preaching on the streets of Minneapolis, Reverend Daniel Hartsough traded his five-passenger automobile for 300 acres of rolling farmland in Barnes County, in the southeast corner of North Dakota. Along with his mechanically minded son, Ralph, he worked their acreage most of the year, while continuing "to exhort the wayward to repentance during the winter months." Together, they conceived a crude, single-cylinder traction engine for use on their farm and returned to Minneapolis to make it a reality.

There, they designed a four-wheel, single-cylinder tractor, followed by two more, an eight-horsepower unit and a twelve-horsepower version (on the belt). They shipped all three to their farm. The eight-horsepower unit "gave a good account of itself," Ralph told the *Implement & Tractor Trade Journal* fifteen years later. "[B]ut I am perfectly willing to admit that the others were bum, very bum."

Over the next few years, they continued to design, experimenting with a three-cylinder version, then a four-cylinder model. By 1906, they had a design they claimed would replace twenty horses and reduce operating costs by four-fifths.[6]

The Hartsoughs filed patents for their design, then sold a half-interest to a Steele, North Dakota, farmer, part-time implement dealer, and part-time thresher manufacturer named Patrick J. Lyons. He had the right temperament to manage the company's affairs. Labor shortages had caused the cost of labor to go "sky high" and made it so the "laborer was practically the boss," Lyons opined, calling to "adopt modern methods else we cannot exist."

"After a thorough inspection and investigation," the trio announced plans to liquidate their property and move to Minneapolis. "I have noticed that in our section of the country," Lyons told a Minneapolis newspaper, "farmers, when contemplating the purchase of machines, always depend upon Minneapolis to provide them."

Their new company, the Transit Thresher Company, contracted with local machine shops to turn out a respectable twenty-five tractors in 1908. Their thirty-five-horsepower tractor was one of five competitors at the inaugural Winnipeg trials in 1908, but failed to place in the top three at the judged competition. "Accidents will happen in the best-regulated families," reported *The Commercial Motor*, "and it was not long before this motor was seen charging the fence due to a broken steerage chain."[7]

Still, Lyons successfully secured two more investors, Fred Glover and John Muir, both North Dakota farmers, and organized the Gas Traction Company. "When we started the Gas Traction Company," he later told a reporter at the *Minneapolis Sunday Tribune*, "we were pioneers and alone in the field of four-cylinder traction engines. Today the entire world is trying to produce an efficient and economical power for operation on the farm."[8]

In 1909, competitors in Winnipeg could partner with the plow manufacturer of their choice. John Deere entered two plows at the event. The largest was a fourteen-inch bottom gang plow, weighing in at more than 3,000 pounds, supplied for the Big Four 30 tractor. Another tractor pulled a smaller seven-bottom version of the plow. And though failing to place, Deere was sufficiently impressed with the Big Four 30 tractor to soon offer it through their network of dealers and agents and, importantly, its newly formed export department, based in New York City.[9]

Deere praised the combined work of the Big Four 30 and the John Deere Engine Gang Plow, one customer calculating that the outfit did the work of forty-eight horses. He offered proof in the form of two photos. One showed a plot of weeds as tall as a man, and the other a "field subdued—the weeds buried—the land lays like a garden."[10]

By the time Deere's representative, Floyd Todd, arrived in Minneapolis in the spring of 1912, the Hartsoughs had moved on from the Gas Traction

Company. Based on their experiences in North Dakota, they wanted to build something even larger, selling their interest and putting the proceeds into a new traction engine capable of pulling a twenty-bottom plow. But orders did not come for their new tractor, the "National," and they sold the single unit to a wheat grower in Montana.

Father and son "almost parted company in their opinion as to the tractor of the future."[11]

In the Hartsoughs' absence, the Gas Traction Company was now assembling fifteen tractors a week in a 43,000-square-foot factory in Minneapolis, with a smaller assembly facility in Winnipeg to support Canadian markets. Sales amounted to just over 450 tractors in 1910 and 1911, with projections for as many as 750 tractors in 1912. The company had also attracted other suitors, according to a Mr. Lucke, general superintendent of the Diamond Iron Works of Minneapolis, the shop contracted to machine most of the parts for the Big Four 30. He was invited to Moline to share what he knew.

A man like Lucke knew more than most, and fully understood that keeping track of the transient stream of entrepreneurs and first-time investors was an enterprise all its own. His Diamond Iron Works was now cutting gears for seven different Minneapolis tractor companies, and through that work could estimate total combined production for the year of 2,800 tractors. The Minneapolis Steel & Machinery Company was building 200 tractors of their own and 500 engines for the J.I. Case company. The Northwest Thresher Company was building 1,000 "Universal" brand tractors. "You remember the Universal tractor was originally built in the Northwest factory at Stillwater," he shared. "They sold out to the Minneapolis Threshing Machine Co., . . . the Minneapolis Threshing Machine Co. have in turn sold the Universal Tractor business to the Rumely Company."[12]

The Gas Traction Company, Lucke shared, had been sold to the J.I Case Threshing Machine Company, though in short time they learned that the buyer was actually the Rockford, Illinois–based Emerson-Brantingham Company. Deere considered the $2 million acquisition an over-valuation but recognized the significance of the transaction. Emerson-Brantingham was in the same race to develop a full line, and the acquisition of the Gas Traction Company provided the capstone of the new $50 million company.

The merger, according to the headline in the *Minneapolis Morning Tribune*, was "formed to battle International Harvester Company."

"We think well of the Gas Traction Company," Deere's Executive Committee recorded, but thought that "the high profits seem of a temporary character." Deere executives thought themselves better served by a more thorough investigation of the market.[13]

Willard Velie's urgency was prescient, and other members of the board began to offer their support. The farm equipment industry, comprised of hundreds of successful regional manufacturers, a few national, and—in Harvester's case—international organizations, was moving uncharacteristically fast. Deere moved more aggressively, and over the next few months, a parade of tractor designers and executives began to arrive in Moline. At the same time engineers began to canvass the country to determine the merits of some of the country's bestselling tractors.

Deere reached out to the Aultman-Taylor Machinery Company in Mansfield, Ohio, in April to determine the "merit of the gas tractor made by this company." The company predated the Civil War, building threshing machines, steam engines, and other agricultural equipment. Their first gasoline tractor, Old Trusty, was comparable to the Big Four 30, which Deere categorized in the "heavy" class and which had just recently started production.

Conversations were held with the Minneapolis Steel & Machinery Company, which began its life as a structural steel manufacturer but had recently embarked on a collaboration with an impressive young inventor named Walter McVicker on the development of gasoline engines for a planned line of Twin City–branded tractors. Those meetings were followed by a review of two tractors called the Imperial and the Cascaden, the latter a yet-unbuilt design that, "if perfected," looked the most promising of all machines seen to date.

In May, Deere joined forces with one of the most successful gasoline tractor manufacturers to date, the Hart-Parr Gasoline Engine Company, for a joint exhibit at the Chicago Land Show, a miniature world's fair with exhibits from companies across the Midwest and West selling everything from clothes to home furnishings to farm tractors. Deere contributed a selection of its "huge gang plows."[14]

In only a few short months of investigation through the spring and summer of 1912, Deere was discouraged by just how fragmented the nascent tractor industry had already become. But a new generation of voices, emerging from a series of promotions essential to shaping the newly consolidated company, began to join Willard Velie's challenge to wholeheartedly enter the market.

Fresh from a six-month investigation into agricultural markets in western Europe and Russia, vice president of manufacturing George W. Mixter joined Velie's aggressive agenda for tractor development. Mixter, the round-faced great-grandson of John Deere, was a third-generation Yale graduate, the son of a chemistry professor. Like his father, he went from Yale to the Sheffield Scientific School. He completed advanced electrical training at Johns Hopkins University, joined the Minneapolis sales branch in 1897, then moved to Moline a year later with the task of developing a cost accounting model for factory inventory. By 1899, he had become a master mechanic at the Plow Works; then, at the turn of the century, he spent ten months in Buenos Aires solving field problems for Deere equipment sold in Argentina, Chile, Uruguay, and Brazil.

Mixter was superintendent of the John Deere Plow Works by 1904, a director in 1905, and vice president in 1907. He was now responsible for the full reorganization of Deere's manufacturing facilities. Understanding Deere's manufacturing capabilities better than anyone, he began to work closely with head of engineering Max Sklovsky, a twelve-year veteran of the company, and lead designer Charles Melvin. They narrowed the growing list of tractor manufacturers to what they considered to be the best of three tractor classes: heavy, small, and motor plow.[15]

In the "small" class they singled out the two-cylinder, 7,500-pound 12-25 tractor with fifty-two-inch rear wheels from the Avery Company of Peoria, Illinois, calling it out as the best tractor in the class. The full Avery lineup included the two-cylinder 8-16, four-cylinder 14-28, 18-36, and 25-50. The Avery 40-80 weighed in at eleven tons.

In the "motor plow" category, they looked to the Hackney Motor Plow and immediately went to see it in person. On May 6, 1912, Sklovsky and Melvin left by train, then secured a car and drove to Warren, Minnesota, to visit a farmer operating a Hackney Auto Plow with three plows in a field of wheat stubble. Farmer Pembina told the Deere engineers he had plowed 175 acres over twelve days and was consuming, on average according to his notes, twenty-five gallons of gasoline for every sixteen acres. He had another 1,500 acres to plow. Pembina operated the plow for them at a depth of eight inches, and everything held up well. The wheels were clogged with dirt, but still the tractor performed well with no engine trouble. Plans for a return trip to follow up on the Auto Plow's "wear, strength, and durability" was noted. The following day they visited one of the Hackney brothers (reports did not name which one) at their factory on University and Prior Avenues, Minneapolis.[16]

Brothers Leslie S. and William L. Hackney got their start buying and reselling land in North Dakota. They turned that into a venture for making hay tools in 1901. After enlarging their factory several times, then relocating, they formed the Hackney Manufacturing Company in 1909, making everything from hay tools to farm gates, many based on their own designs. This led to the development of an ingenious design for "Combination Tractor and Farm Machine," for a variety of on- and off-the-farm work, from plowing to seeding to harvesting to road grading. It was a three-wheel design with a 22-40-horsepower rating, three forward speeds, and a power lift system, a mechanical method of raising and lowering implements available on horse-drawn implements for decades. The Hackney came standard with three under-mounted plows. Two opposing seats allowed the operator to run the tractor in either direction—one for plowing and one for grading.[17]

Manufacturing had only started in August 1911, but by April 1912 they had already sold 100 tractors and employed twenty-two workmen and three stenographers. They forecast total sales of 350 machines in their first full year.[18]

One of the Hackney brothers, their sales manager, and their head engineer hosted the group from Deere. Fifteen tractors were being built, and they claimed to be finishing one a day, "which seems quite plausible after what we have seen at the works," Melvin reported.

They watched a demonstration of a test tractor, "an unfinished, rough affair used in all their experimental work," at the Minnesota State Fairgrounds in St. Paul. Melvin knew the area territory well because of his first job with the Minneapolis Harvesting Works, which he left for Deere in 1900. At the fairgrounds, design deficiencies began to emerge. Melvin learned more from one of Hackney's mechanical engineers, a young man who happened to be boarding at the same house as Melvin's son. The Hackney brothers were experimenting with a new model, he told his roommate's father, but were not ready to share.

Fortuitously, on the return train to Moline, Melvin talked with a farmer who had himself just visited the Hackney factory. He had been farming with horses for twenty-eight years, he shared, and now "was convinced in his own mind" he would buy a tractor for his 350-acre farm.

On his return to Moline, Melvin advocated for the "merits of the Hackney Motor Plow," and in July 1912, was transferred from the Plow Works Experimental Department for the purpose of designing and building an experimental tractor. His $6,000 budget, when an average wage was 22 cents an hour, was quickly consumed in the production of the first experimental John Deere tractor.

Melvin's tractor, which admittedly bore "similarity to the design of the Hackney Motor Plow," featured a three-bottom plow mounted underneath the chassis in between two front wheels and a single, smaller rear wheel, though in fact there was no front or rear. A platform behind the engine and large wheels included two seats with a single vertically mounted steering wheel and single set of control levels. For plowing, the operator sat facing the motor, using a crank to lower the plow into the ground. For other operations, the drawbar was mounted in front of the engine, below the radiator. The operator moved to the other seat, with the engine and whatever was being pulled behind him.

Charles Melvin's experimental motor plow, 1912. Courtesy John Deere.

The "results were disappointing both as to field performance and keeping the tractor from breaking down." The tractor was built mostly from existing parts and operated primarily as a plowing tractor. A production version had to be less costly to build, and more affordable and versatile. As quickly as it was built, the Melvin design was canceled. Still, it was a critical beginning, even if it failed to resolve the carefully worded board resolution of March 5, which set a clear objective to divorce the tractor from the plow. Fortunately, a new design was already underway.[19]

5

"THE GREAT AWAKENING"

On June 28, 1914, Austrian Archduke Franz Ferdinand was assassinated, triggering a series of ultimatums across Europe. The Duke was traveling in a motorcade, fully exposed in a topless automobile. Germany declared war against Russia on August 1, followed by declarations of war from France and Great Britain. Over the next four years of unprecedented carnage, a new era of technology, facilitated by war, was adopted and embraced, proving again and again that necessity was truly the mother of invention.

The world was much smaller than it had been even a decade before. In 1909, Deere's George Mixter spent four months demonstrating Deere plows in Russia and studying the potential market there. He reported an existing market for a million plows a year, "mostly very poor articles," but for other implements from America there was "practically nothing" except for a presence by the International Harvester Company. In Russia, he reported, Harvester was even "stronger than in U.S.A."

Russia's soil and climate were both favorable, and the market was large, Mixter reported. But the "large expenditure necessary to create a plow trade in

Russia should not be undertaken, unless it be coupled with a determination to ultimately manufacture in Russia," which he estimated would cost $150,000 to get even a modest factory running. The "enormous possibilities" could not be ignored, he submitted, but, somewhat prophetically, he said that the country was politically unstable, and any investment could be quickly lost.[1]

International Harvester was uniquely exposed, the efforts of general manager Alexander Legge now netting half of the company's sales from outside of North America. Recent expansions of twine mills in Neus, Germany, and Norrkoping, Sweden, a new malleable foundry and second twine mill in St. Croix, France, and a new warehouse and equipment for the production of grain binders in Lubertzy, Russia, were now all at risk.

Americans read about the war in newspaper dailies. Few yet had radios. President Wilson continued to assure that America would remain neutral, and in most ways, life went on without interruption. Despite the frantic pace at which farm equipment manufacturers were reorganizing, consolidating, and introducing new equipment forms, farm life had changed very little for most. Farmers still relied on horses and horse-drawn equipment to plow and till their fields, and then to cut, gather, and process their crops. Thousands of enormous steam traction engines continued to work the western prairies, but now only a few companies remained to meet the dwindling demand. The number of gasoline (some of them ran on kerosene) tractor manufacturers had doubled between 1912 and 1914, now thirty-nine in all, but only a handful were responsible for the bulk of annual production that now reached a record 11,000 tractors.

In 1913, the final year of the Winnipeg Motor Competition, the American Society of Agricultural Engineers volunteered to judge any motor contest organized in the United States. Their objective was the development of standards.

The First Annual National Power Farming Demonstration, sponsored by the *20th Century Farmer* magazine, took place in September on the Coad farm, just northeast of Fremont, Nebraska. Ten thousand farmers watched eighteen tractors work 160 acres of land at one time. Other reports claimed forty tractors on 500 acres.[2]

Not only did the modest show in Fremont draw a small crowd, but it fueled an inevitable competition against the tractor's greatest adversary, the horse.

Organizers of the next National Power Farming Demonstration expected to "deal 'horse farming' another blow that will push it just another step into the oblivion which it is one day to occupy beside the old ox farming of a half-century ago." Farmers were not yet convinced.[3]

Lynn Ellis, author and analyst from the United States Department of Agriculture, doubted the horse would be cast aside anytime soon, exploring the "problem of the small farm tractor" for *Scientific American* in 1913. Manufacturers had yet to overcome the fact that "the horse is by far the most flexible motor for small jobs." Challenges of mechanical power for pulling plows and cultivators, hauling empty trailers and trailers with loads, and field issues like plow bottoms hitting large rocks and stalling or damaging the engine were serious challenges to overcome, he rightly assessed. Tractor width, fuel type, weight, and myriad other issues were still hindrances to the purchase of a small tractor.

"As yet the small farmer cannot supplant all his horses with a small tractor," Ellis wrote. That meant that, at best, the tractor offered supplemental power to replace a few horses on the farm. The horse was not going anywhere.[4]

Henry Ford's public pronouncements on tractor development had gone silent while the Model T captivated America. Ford sold 20,700 Model Ts in 1910, 53,500 in 1911, and a staggering 82,400 automobiles in 1912. His assembly line, the innovation that put him on the road to worldwide adulation, was put to full application in June 1913, and with astounding results. The Ford Motor Company built 185,000 automobiles over the next year. "If there were no Fords, automobiles would be like yachting—the sport of rich men," went an advertisement for the Model T. But the Ford "brought the price down within reason—and the easy reach of the many."[5]

Henry Ford's new Highland Park factory forever upended industrial manufacturing by abandoning the concept of compartmentalization, and instead concentrated production under one roof and within a specific sequence of construction. Assembly now informed design, and each time a competitor inched closer, Ford seemed to quickly put greater distance between them. Industry

A young farm family in their Ford Model T, 1910. Image from the Collections of The Henry Ford.

sales of commercial vehicles, which included automobiles and trucks, exceeded 700,000 in 1915, valued at more than $500 million. Ford outsold his closest competitor, the costlier Willys Overland, by more than five to one.

Improved roads, new bridges, and filling stations were in high demand, essential infrastructure to facilitate a growing appetite for travel and adventure. New fashions evolved, as did redesigned luggage, camping equipment, picnic baskets—everything reimagined to be packed in a vehicle for Americans who were bombarded with messaging to "see America." A new class of transcontinental tourists sent letters to local newspapers asking for routes. Yellowstone National Park opened for automobile traffic. Sadly, the Higgins family, of Elliston, Montana, skidded off a park cliff, killing Mrs. Higgins. Mr. Higgins fractured both legs, but he and their two children survived.

In Kansas, a crime ring ran a scheme to fill mudholes each night with water. When unsuspecting motorists became stuck, a group of men conveniently stopped with their tools, charging $5 for five minutes of work. Farmers in Missouri, whom the state highway commissioner dubbed as "mudhole pirates," were accused of the same racket.

Ford's Highland Park factory was perpetually under construction in a race to meet demand, while ancillary departments to manage all aspects of work and life sprang up. The new sociological department at the Ford Motor Company began teaching English and other courses to help Americanize its growing immigrant workforce. Half of the factory wage earners at Ford spoke no English. Housing developments, schools, churches, and hospitals sprang up to support the Ford factory.[6]

❦ ❦ ❦

In rural America, "the great awakening came in 1914—fired by a spark from an entirely unexpected source," the *Implement and Hardware Trade Journal* recalled fifteen years later, not about the world going to war, but about the arrival of "a small farm tractor announced from Minneapolis." The tractor was named the Bull, and the company was simply named the Bull Tractor Company. Both the tractor and the company were new, but the men behind the scenes were anything but newcomers.

A few years removed from their withdrawal from the Gas Traction Company and the subsequent multi-million-dollar sale of it (from which they earned no proceeds), Reverend Daniel Hartsough, Ralph Hartsough, and P.J. Lyons, the latter profiting handsomely from both deals, reunited on a new design. Father had finally given in to his son's wishes to build a tractor for the average farmer.

Similar to their previous arrangement, the tractor was designed by the Hartsoughs, financed by Lyons, and built by a third party, in this case the Minneapolis Steel and Machinery Company. A separate company was organized to build the engine, fittingly named the Toro Manufacturing Company.[7]

The Bull was the affordable tractor that Henry Ford planned to deliver. Unlike the strong, muscular animal for which it was named, the tractor was long, close to the ground, with a triangular stance, two rear wheels, and a single front wheel. Offering only five horsepower of pulling power at the drawbar, it claimed ability to pull two fourteen-inch plow bottoms, perfect for the average-sized farm. Perhaps most importantly, it was simple to understand and operate and was inexpensive, selling for a mere $395.

The Bull Tractor Company burst onto the scene with the $395 Bull tractor. Courtesy John Deere.

"It was never a mechanically sound product," Harvester's Cyrus McCormick, Jr., later reflected on the Bull's impact, "but its commercial popularity was such that it swept the field."

In Omaha, the Bullock Machine and Supply Company bought 750 Bull tractors to sell in Nebraska and western Iowa, supplementing tractor sales with the Sanford auto truck and an automobile called the Ohio, which, although new, was "considered among the really good cars." Bullock wanted more, but the Bull Tractor Company could not build them fast enough.[8]

The Bull single-handedly accelerated industry sales, which reached somewhere between 10,000 and 16,000 tractors nationwide—no process had yet been established to track sales, so numbers were useful to track trends—but offered little in terms of power ranges or market share. It was clear, though, that the Bull cornered the market overnight. Tractor sales at International Harvester, which had grown steadily year over year, an expected trajectory for a field-tested and properly marketed product, plummeted, falling 76 percent, to barely more than one thousand tractors in 1914.[9]

"The gorgeous nightmare period of the industry followed," an *Implement and Hardware Trade Journal* reporter wrote to describe the period to come.[10]

In Huron, South Dakota, A.M. Urquhart, a Ford distributor, advertised an economical pair, a Ford car and a Bull tractor, the latter with a massive mark-up.[11]

> *A Ford Car Costs you $530 F.O.B. Huron*
> *A Bull Tractor costs $625 F.O.B. Huron*
> *Total for two $1155.*
> *The Ford does all that a big car will do and more*
> *The Bull Tractor does the work of seven horses and at less expense*
> *You can have the two for the price of a large car*
> *ENOUGH SAID*[12]

The Model T catapulted Henry Ford to national fame, a doting American public devouring news of even the most trivial activities of the man from Detroit. In May 1915, Ford, through advertising manager C.A. Brownell, at last revealed his long-awaited plan for the farm tractor, though the tractor itself had not yet arrived. "The new tractor will be a wonderful machine," Brownell told a room of area automobile dealers in Des Moines, Iowa. "It is the hope of Mr. Ford that he will be able to place it on the market at about $200. It will be ready for introduction about December 1." A prominent "feature" was that every Ford serviceman would be available to service the tractor. "The engine," he stated, after touting three years of development, "will probably be the same as installed in Ford cars."[13]

The tractor would have the "typical Ford front appearance," went another report, and "will be of such a reasonable price that practically every farmer in the country will be able to buy one." It confirmed the $200 price tag, "although it may be a little more." It would do the same work as six horses.[14]

Internally, Ford's tractor faced stiff resistance from shareholders, particularly John and Horace Dodge. The Dodge brothers left Ford's employ in 1913 to form their own company, which was financed by dividends from their large holdings of Ford stock. Other shareholders had entertained Ford's tractor hobby so far, but their willingness to overlook the funneling of monies into a speculative, and certainly unprofitable, tractor endeavor had ended. Ford's introduction of a $50 rebate to all Model T owners, which cost the company more than $15 million, didn't help.[15]

To avoid further delay, Ford simply moved forward without them.

In November 1915, a vague yet promising announcement offered plans for the formation of an independent company and construction of a dedicated tractor factory. It would be managed by his twenty-two-year-old son, Edsel, with Henry investing up to $1 million of his own money. "In the new tractor plant there will be no stockholders, no directors, no absentee owners, no parasites, declared Henry Ford the other day in a discussion of modern industrialism. There will be no incorporation. Every man employed during the period of his employment will share in the profits of the industry."[16]

In fact, developmental work on the tractor, which had slowed but never fully stopped, was picked up again by Joseph Galamb, the draftsman who first impressed his boss with work on the Model A, then as chief design engineer on the "universal car," the Model T. In a carriage barn not far from Ford's original Piquette Avenue factory, Ford, Galamb, and a few others built three new tractors in 1914 and 1915, adapted from an older Ford Model B and the more recent Model T.[17]

Eugene Farkas, a Hungarian immigrant now in his third and final stint at the Ford Motor Company, joined the tractor team as well. "I knew seventeen words" in English, he later said about his arrival to the United States in 1906, "and you'd be surprised at how you can get along with so small a vocabulary."

By the time Galamb convinced Ford to hire Farkas for a third time, in 1913, he had worked for no less than eight automobile companies in and around Detroit. As chief of the experimental drafting room, he was earning the princely salary of $165 a month.[18]

Ford asked Charles Sorensen, the supervisor of the pattern department and assistant production manager at the Piquette Avenue factory, to construct a new room on the third floor for use on a new project. Ford picked Sorensen to, as Farkas put it, "take over this new tractor plant that wasn't built yet, to manufacture the new tractor that wasn't designed yet." When they arrived in

October 1915, there were floors in the pattern shop, window openings with no windows, and the pattern makers were building their own benches—they had nothing yet to do and no furniture on which to do it. They had installed screens in the windows in a failed attempt to keep the flies out—"we were just killing flies half of the time and trying to work at the same time."[19]

In the design room, Galamb and Farkas erected a large blackboard, and Ford, always the sentimentalist, hauled in his mother's rocking chair "for good luck." There, rocking back and forth, he spent hours carefully studying their work and asking questions. "He was right in there changing things that he didn't like."

Prints were made, sent to the pattern shop to create the patterns, and then sent off to assembly. The first tractor was finished just after the first snow fell in December 1915, just enough time to get into the field and plow.[20]

6

"SMALL TRACTOR PROPOSITION"

Only 15 percent of American land was under cultivation in 1850, and with each passing decade, population growth accelerated, creating correspondingly greater demands on agricultural production. Between 1910 and 1920 alone, the population had grown nearly 15 percent, to more than 105 million people. There were 25 million horses on 6.5 million farms in the United States. Half of the country's population still lived on the farm, or at least were categorized as rural dwellers, but that number was shrinking enough to cause concern in agricultural circles. In particular, young people were being seduced by the city. Agricultural equipment manufacturers recognized the difficult battle ahead to keep them down on the farm.[1]

"A generation ago farming was an occupation. Today it is a profession—a science," offered Walter Remley of the International Harvester Company in a radio address, part of a series that put the voices of educators, executives, and others in front of a growing audience of radio listeners. This talk in particular, "Why and How the Farm Boy Will Beat Dad's Time," pleaded with the "farm boy" to stay on the farm, where opportunities were more diverse and independent and more sustainable than unskilled factory work in the city.

"Civilization depends more on the farmer than it does on any other business or professional man," Remley deduced. It was the industry's latest effort to convince the younger generation to stay home and embrace power farming. A new era in the technology and science of farming was just beginning, passing from "the pioneer stage of farming into something."[2]

Farmers were early technology adopters. Small gasoline engines, many of the one-and-a-half- to three-horsepower variety, had become commonplace as a source of power, and, more importantly, the source of additional income. The gasoline engine, according to an International Harvester Company advertisement, gave "your wife a chance to develop into a business woman." A two-horsepower gasoline engine and cream separator, for a mere two cents an hour, ran the churn, the washing machine, and cider mill; drew water; sawed wood; chopped feed; and shelled and shredded corn—for starters. While saving the time of a farm hand to look after the cows, it empowered a farm wife to "quadruple her earnings from butter, eggs, [and] chickens." Nothing was wasted on the farm, and the gasoline engine offered greater earning power and "relieves the womenfolk of drudgery."[3]

Farmers were also better educated than ever. General trade journals like the *Farm Implement News*, *Twentieth Century Farmer*, *Motor Age*, *Farm Journal*, the *Country Gentleman*, and the *Prairie Farmer* kept readers informed of weather, new farming techniques, industry consolidations, and company and personnel news. Manufacturers had their own voice as well. In 1895, John Deere introduced *The Furrow*, a journal for "progressive farmers." By 1912, *The Furrow* was being printed in two colors and reached four million farmers. By 1916, it would expand to seventeen different regional editions.

It was a statistical inevitability that farmers would buy more automobiles than anyone, and they were buying them faster too. Conversion kits could also now turn a car into a vehicle for almost any use. In time, tractor conversion kits were available from Adapt-O-Tractor, Smith-Form-A-Tractor, and dozens of other companies. Cheap transportation and newly paved roads created new opportunities for farmers to market and to sell surplus crops.

⁂

The automobile sold Americans on the benefits of leisure, but farm tractors had to generate profit within the boundaries of existing farms, most of which were smaller than fifty acres. Incremental gains from investments in new, less expensive agricultural implements offered savings in both time and labor, though whether those savings were causing young people to leave the farm or instead facilitating greater work in their absence, was hard to determine. Tractor adoption, which required a heavier up-front investment and a redesign of time-tested daily operations, added new complexities and risks to an already risky profession.

Selling a suitable tractor was only one part of the equation for long-standing relationships among manufacturers, dealers, and customers.

At John Deere, tractor work was divided among Max Sklovsky, Joseph Dain, and Theo Brown, who were asked to "do some thinking about tractors" during the summer of 1915. "My work is not directed by anyone," Sklovsky wrote shortly after joining Deere, somewhat perplexed, "and the superintendent has told me to go ahead, using my own judgment about most things." Later he would attribute his success to having "got in with good people, and they pushed me along."[14]

A successful design, it was clear, was only part of the equation. Farmers eager to adopt the small tractor were quick to discover that advertising did not always accurately reflect performance. After its meteoric rise, an equally swift demise had already started for the Bull Tractor Company, a not uncommon trend for the now hundreds of entrepreneurs clamoring to enter the tractor business. The introduction of the Big Bull 12-25 in 1915 offered horsepower improvements over the original machine, now renamed the Little Bull 5-12, but farmers quickly discovered the limitations of each model. Farmers pushed the tractors to perform additional operations for which the machines were unsuited. The Bull could tip sideways doing work on hills, or backward under the strain of a plow. Deaths were reported in a few cases, and the dangers of tractor operation became a too-often-discussed topic at grain elevators, dealerships, and county fairs.

The Bull Tractor Company's partners began to fight over its future, Reverend Hartsough finding fault with partner P.J. Lyons, the financier and general manager. A long, drawn-out intellectual property battle ensued. In 1916, a federal court sorted through the complicated relationship, which collectively included at one time or another the Gas Traction Company, Bull Tractor Company, the Hartsough Tractor Company, and included tractors called the Little National, Lion, Lyon, Bull, and Big Four 30. The ongoing litigation would slowly kill the Bull over the next few years, but its low price continued to prove attractive, despite its limitations.

Soon after Deere tested its first experimental tractor, built by Charles Melvin in 1912, George Mixter began to lobby for more permanent development space and to "transfer the manufacture of some additional lines to East Moline or to take on the manufacture of other lines, especially light tractors . . ."

Deere turned to Joseph Dain to design a light tractor they could sell for $700. Through the fall and winter of 1914, Deere spent another $3,000 on development of a new prototype to complement the work of others already underway.

Mixter authored a brief history of the John Deere Marseilles Company and its progress through 1914, a task he was completing for all of Deere's recent acquisitions. He used the history to argue that "the capacity today is greatly in excess of the demands, due primarily to the greatly increased efficiency of producing the product." The factory could utilize the foundry of the John Deere Harvester Works next door, transfer the blacksmith shop to the existing foundry space, and use the blacksmith shop, "the cheapest of sheet metal construction," and the most "excellent space for the erection of small tractors . . ."[5]

Mirroring common conversation throughout the industry, Butterworth and the executive board continued to debate the merits of tractor development. In correspondence in the summer of 1915, Mixter outlined the most promising scenarios, pointing to how the country was being "flooded with attempts at practical small tractors" and that the "farmers are right in demanding a small

Joseph Dain's first prototype, in February 1915, of what became the All-Wheel Drive. Courtesy John Deere.

tractor." None of them, he contended, were "built with the proper spirit behind the design and manufacture to insure their durability in the hands of farmers."

Growing automobile sales continued to feed expectations that tractors would operate as easily as a Model T or Chevrolet's Baby Grand, now available with an optional electric start engine, an alternative to the traditional hand crank start. Owners could troubleshoot most problems and often make repairs on their own. And with a wide array of automobiles now available for less than $1,000, price made them accessible to most.

Mixter dismissed the "big tractor outfits" as "things of the past" and outlined a series of scenarios by which Deere could successfully enter the tractor business.

1. Form a new tractor company with stockholders to subsidize the business.
2. Contract an existing supplier to manufacture the full machine.
3. Contract components such as the engine and possibly assembly, etc.
4. Manufacturing entirely by Deere & Company.

Mixter advocated for outsourcing as much of the tractor as possible, which would concentrate efforts more on advertising and less on manufacturing, he argued. "It seems to me probably that one or two years from now if we are successful in conducting Deere & Company's business as to show up well," he wrote Butterworth, "that our principal banker friends might be consulted as to the wisdom of a small tractor proposition and then at that time it might not look as undesirable as it very rightly has looked to you and to the bankers during the past year."[6]

Deere's Joseph Dain built three more versions of what was an all-wheel, chain-drive concept, in 1915. The first weighed 3,800 pounds, twice the weight of a Ford Model T, and half the weight of the Avery 12-25 tractor. A simulated steady drawbar pull of 5,000 pounds was achieved on the slowest transmission speed, though field tests pulling three 12-inch plows at a speed of two and a half miles per hour achieved only 3,000 pounds of pulling power. In further field tests, the chains proved too light, and two front ratchets broke. Undeterred, Dain was pleased overall, noting that "as it is entirely different from any other tractor on the market, we did not have anyone's previous experience to guide us."

The second Dain shared the first's friction transmission, a common but inefficient system that transferred converted energy from two transmission wheels to the tractor's drive wheels. The new machine weighed 4,000 pounds. On a Minnesota farm, it plowed eighty acres at 59 cents per acre, inclusive of one man's time at 30 cents per hour. Neighbors watched while the tractor effortlessly pulled three fourteen-inch plow bottoms. The ground was too wet for their six-horse team to pull their green and yellow New Deere gang plow.

Rapid progress continued, and the third Dain prototype was completed shortly after Thanksgiving 1915, just a month before Henry Ford's team was putting its latest experimental tractor into the field for tests. Dain replaced the friction transmission, which was not uncommon in contemporary designs, but lacking in power for plow work, with a positive gear-driven type, a mechanical solution that prevented slippage. He was enthusiastic—on March 13, 1916, he dispatched a telegram to Moline from the test site in San Antonio, Texas.

> *Have followed tractor closely for two weeks. Conditions extremely hard and rough. Absolutely no weakness in construction. Gears, chains, universals, in fact all parts in good condition. Tractor has traveled near five hundred miles under extreme load. Change speed gear a wonder. I recommend to the Board that we build ten machines at once.*

An additional $25,000 to $50,000 was required for new machinery, patterns, and tools for full production, he enthusiastically told the board. Approval soon came for building ten more machines at the John Deere Marseilles factory in East Moline.[7]

In the summer of 1916, Deere enlisted the services of Walter McVicker, a mechanical and electrical engineer from Michigan, to design an entirely new engine in an effort to overcome power issues with the Waukesha engine currently in use on the Dain tractor. Over the next decade, the Waukesha would power nearly 150 different tractors, but it was the wrong one for Dain's tractor. McVicker was a prolific inventor, perfectly suited for the age. He designed for power and speed. In 1903, he had patented a gearless engine, the McVicker Automatic Gas & Gasoline Engine, built and sold for the next decade by the Alma Manufacturing Company in Alma, Michigan. The gearless engine had one-third the parts of contemporary designs. More patents followed.

A consolidation of the Alma Motor Company and Alma Manufacturing Company in 1904 led the way for the manufacture of McVicker engines, with development of a grain-loading machine progressing, in addition to plans for trucks and automobiles. By 1911, they were selling a full line of stationary engines and the Cameron Auto Truck, and in January alone, they built more than 1,000 "Alma Junior" engines, their latest offering.

In 1907, McVicker tested his first high-speed railroad car, capable of a top speed of forty-five miles per hour. An automobile patent followed.

Always price-conscious, William Butterworth worried about the most recent cost analysis of the Dain tractor. Henry Ford claimed that his tractor, when available, would sell for $200 complete. The new McVicker engine alone would cost Deere $200 each, with full machine estimates now expected to run at least $600, and a projected retail price of $1,200, nearly double the target sales price of $700. "This is somewhat higher than has been considered admissible for a three-plow tractor," Dain admitted. "It is the writer's belief, however, that an all-wheel drive will ultimately be the tractor the farmer will pay for."

Meanwhile, Max Sklovsky was nearly ready with the second iteration of his two-plow tractor under the experimental number B-2, now described as a "small edition of the Dain machine." A pivot-axle and automobile-type steering were added, which, like his first design, featured a four-cylinder Northway engine built by a subsidiary of General Motors. Two updated versions of the Dain, one with a Waukesha engine and one with the new McVicker engine, were now being built.

Attempts to consolidate the projects failed for the time, Brown confiding to his journal that "It is going to be quite a proposition for George Mixter to get

Max Sklovsky at the wheel of his experimental B-2 tractor, 1916. Courtesy John Deere.

Dain and Sklovsky together on a tractor design. Max is extremely obstinate and I think altogether to[o] much so and it is a real proposition to solve."[8]

Dain was still fighting for the all-wheel-drive concept, overlooking the revelations that had come from sales of the Bull tractor and the expected path of the Ford. "It is probably true that the inclination of the farmer is to buy a tractor that is as much like an automobile as possible," he told the board. "On the other hand," he rationalized, "offering an all-wheel-drive machine might overcome this point."[9]

Mixter was convinced that customers wanted a tractor that steered like an automobile, but Max Sklovksy's tractor design, which included automotive-type steering, offered still a "question in the writer's mind of the commercial practicability [*sic*] of a high-grade four-cylinder machine for two-plow work."[10]

Joseph Dain proved more convincing, and Sklovsky scrapped his four-cylinder tractor and drew up a third concept, the D-2. It shared the same single-piece, cast iron frame concept as the first two but now featured a single-cylinder engine. He also abandoned the automobile steering to align with Dain's all-wheel, chain-drive design, finding some middle ground with Dain after all. "Low cost, easy accessibility, certainty of burning kerosene being some of the desirable features," wrote Mixter.

Deere's engineers were making progress, but the industry, driven by world events, was moving faster than ever around them.

7

HENRY FORD DAY

By 1916, farmers, reporters, and equipment manufacturers saw glimpses of just how far the farm had come in visits to their local dealers or at their county fairs. But for one week each year, Fremont, Nebraska, became the epicenter of the agricultural universe, where progress in all shapes, sizes, and horsepower ranges could be found.

Named after American explorer and politician John C. Fremont, the Pathfinder, his namesake town, Fremont, was perfectly situated just northwest of Omaha, less than sixty miles northeast of Lincoln, the state capital. Its central location made it a hub in many senses of the word. The first transcontinental telegraph line came through Fremont in the 1860s, followed by the Union Pacific, then the Sioux City and Pacific Railroad.

Although the "people are industrious, wide-awake and energetic," according to an 1852 travel guide, it was built as a town to pass through. Fremont was along the path of the Mormon trail, the 1,300-mile trek that Mormon leaders Joseph Smith, Brigham Young, and their followers took from Nauvoo,

Illinois, to Salt Lake City, Utah, on their escape from religious persecution in the 1840s, 1850s, and 1860s.[1]

Fremont, the "hub of a sight-seeing tour of Eastern Nebraska," was still growing, a "population of 20,000 by 1920" according to the slogan of the Commercial Club, the lead local promoter of the farming demonstration. Eight grade schools fed Fremont High School, proudly graduating seventy-six students in 1916, twenty more than the previous year. Even the Catholics had a "good parochial school." The Methodists claimed the largest church, with more than 850 members.[2]

Twenty-eight tractors came in 1915, working a larger site just west of Fremont off the Lincoln Highway, the recently completed transcontinental route that connected New York City to San Francisco. Each year the entries and the crowds grew. And for the first time, parking was now an issue. All four sides of the showgrounds were surrounded by all types of automobiles branded Ford, Willys, Chevrolet, Auburn, Buick, or Velie. Or perhaps they came in one built by one of the many agricultural brands such as International Harvester or the J.I. Case Threshing Machine Company. One man drove his Oldsmobile Eight, a lightweight, eight-cylinder automobile, from Fremont to Lincoln and made it twenty miles on one gallon of gasoline. Economy was newsworthy.[3]

They parked as far as a mile away. Locals provided taxi services.[4]

"The third year, 1915, the tractor demonstration 'bug' had so infected the minds of manufacturers, farmers, and farm papers that tractor demonstrations grew overnight all over the country like magic," wrote J.B. Bartholomew of the Avery Company, a Peoria, Illinois–based manufacturer, and chairman of the Manufacturer's National Demonstration Committee. Smaller demonstrations were popping up across the country and were already becoming "uncontrollable."[5]

In the pages of *Gas Power* magazine, Bartholomew beamed that the shows for 1916 would be "unquestionably the greatest agricultural events ever held." He had said the same thing the previous year, and the year before that; so far, he had not been wrong.[6]

The 1916 circuit kicked off in Dallas on July 18 and ended in Madison, Wisconsin, after three months and eight cities. Many considered the shows a publicity stunt, a freak show of the Barnum & Bailey variety, or akin to a traveling carnival. On a daily basis, the claim was hard to dispute as the festival atmosphere, complete with company mascots, Victrolas, motion pictures, nightly hog roasts, musical performances, and throngs of spectators and media filled in the hours of downtime between product demonstrations.

Organizers estimated expenditures of at least $750,000 on the events, some of which went toward the seventy-five train cars loaded with equipment, people, and supplies from sixty-four tractor companies and nineteen plow manufacturers. The traveling agricultural exhibition journeyed from Dallas to Hutchinson, Kansas, to St. Louis to Fremont, Nebraska, to Cedar Rapids, Iowa, to Bloomington, Illinois, to Indianapolis, before finally reaching Madison, Wisconsin, in early September.[7]

But the featured stop for 1916, the fourth of eight, was Fremont. There, Henry Ford would make his long-awaited debut. So would his tractor.

International Harvester president Cyrus McCormick, Jr., who had recovered from a hit-and-run car accident earlier in the year that left him with severe face and arm lacerations, led the army of company representatives that converged in Fremont. J.A. Everson, manager of Harvester's tractor business in Chicago; E.M. Bradley, F.W. Lewis, and R.M. Stewart from the Omaha and Council Bluffs offices; and a fleet of salesmen, machine operators, and others from their Chicago headquarters spent the week there, too.[8]

McCormick, always focused on the company's return on investment, was optimistic. "A collective exhibit like this impresses one with the force and the future value of the industry more than would any scores of individual exhibits," he told *The Bee* out of Omaha. The demonstrations attested that the event was not "pleasure gathering only," but of great value "to the manufacturers and sellers of power, and to the farmers who are purchasers of power."[9]

Sales staff and dealer representatives were plentiful. They represented companies in the field and were the link between manufacturers, who sold their equipment wholesale, and customers. In many cities, typically located near rivers and railroad centers, you could cross the street to visit a competitor dealer and then move on to the next if desired. In the case of Harvester, which was formed in 1902 through the consolidation of McCormick and Deering, you could visit one, or both, of the lines of the parent company.

International Harvester was accustomed to being the featured attraction, not only for its wide equipment offerings, but also for its attractive signage, inviting atmosphere, and the noticeably larger footprint of its operation, anchored by the company's full line. In Fremont, Harvester featured two tractor lines, the Mogul and Titan, which were "shown exhibiting their prowess at all manner of farm power operations."[10] The introduction of the Bull created a temporary lapse in sales for Harvester, but they quickly rebounded with sales of nearly 5,000 tractors in 1915. That number would grow to almost 12,000 in 1916, 40 percent of the industry total.

Harvester brought five models to Fremont: the Mogul 8-16, Titan 10-20, Titan 15-20, Titan 30-60, and Mogul 12-25 tractors. The Mogul line was available exclusively from McCormick dealers, while the Titan could be found at the Deering dealer, usually not far away.[11]

The International Harvester lineup at the National Tractor Demonstration in Fremont, 1916. Wisconsin Historical Society, WHS-46222.

Visitors wandered through Fremont's canvas city, erected by H.W. Rodgers, Jr., the "tent man" from Rogers Tents and Awnings. His Fremont-based company furnished all the canvas coverings for the full National Tractor Demonstration circuit for the year. Electric lighting and telephone circuits were strung along the two primary thoroughfares, Implement Avenue and Gasoline Street. Wallie, the precocious bear cub trolling around the J.I. Case Plow Works tent, drew attention to the company's partnership with the Wallis Cub tractor. Wallie captured more than her share of headlines, at one point escaping her handlers and climbing the tallest tree in Fremont, on Maple Street.[12]

Henry Ford, whose every movement was tracked in newspapers across the country, stopped in Kansas City on his way north to Fremont. There was a national demand for 10 million tractors, he said, and he was going to bundle a Ford car, truck, and tractor for $600. "I'm going to do it," he said, "if I don't croak first."[13]

Ford and a contingent of twenty-five people, including his son Edsel, arrived at Fremont's Union Station, a red brick building with a squared, three-story-tall clock tower, on Sunday at 11:30 AM. A motion picture camera was set up to capture his departure from the train, but he ducked around the corner and into an automobile before being spotted. Sales prospects for his tractor, based exclusively from the national takeover by the Model T, was enough to convince farmers of its merit—or at least its affordability.

Ford made it clear he was on vacation and planned to enjoy Fremont's scenery. Accommodations were provided by the Fremont Commercial Club and its president, State Senator George Wolz, whose hobbies included "good roads, boys and girls scout work, and civic improvement," on the Platte River in a row of houses "handsomely finished." Security guarded the Hormel Bridge to keep out admirers and the press. A six-piece Hawaiian orchestra, brought by Ford, spent the week, and in the evening performed at the local high school. A chef from Omaha cooked for the group.[14]

Though the larger contingent stayed the week, Ford spent only three days in town. Naturally, the press obsessed about Ford, often failing to even mention his team and its introduction of a new Ford tractor. "I have not been able yet to get a tractor that just suits me," he did share, "for I want to make it so that it will work every minute of the time and never fail."[15]

Henry's older brother, William, operated the tractors, which had not yet been given a name or number designation. The board of directors at the Ford Motor Company had blocked its founder from putting his name on the tractor.

At one point William convinced two young girls, Ruth Jardine and Helen Sprague, to "drive his tractors in a private demonstration before the moving picture camera."[16]

Competitors flocked to the Ford tractor, working to secure any intelligence they could on its design, intended release date, or price. Deere's Theo Brown, best known in industry circles for an innovative manure spreader with a "beater on the axle," and factory superintendent George Mixter finally got close. "I got two pictures" wrote the star-struck Brown, "one of Ford, one of his tractor."[17]

Finally, after days of anticipation, the showcase culminated at noon on Wednesday, August 9, 1916. Newspapers estimated a crowd of 60,000 people over the course of the four-day event in Fremont. Others reported 90,000, nearly ten times Fremont's population. A year later the organizers claimed 50,000 on Wednesday alone—Henry Ford Day.

After days of walking and talking, listening to knowledgeable salesmen and company representatives tout the advantages of their machines over the

Spectators arrive for Henry Ford Day at the National Tractor Demonstration, August 9, 1916. Library of Congress, 2007663497.

others, the most dedicated returned for the true test, the plowing demonstration. A dry, dusty haze added mystery to the excitement as, "above the hum of voices arising from the black masses of spectators could be heard the deep-throated exhausts of giant engines idling lazily." Lined up, single file, ready to swarm across hundreds of acres, eighty tractors opened their "powerful steel lungs," reported the *Farm Implement News* soon after. The crowd, at least many among them, had never seen a tractor before the event.

A few hours later, it was over. Tractors returned to their allotted locations within Tent City. Some arrived on their own power, others were towed, while some remained stuck in the field, overheated or worn down from the work.

The agricultural media largely downplayed the appearance of Henry Ford's tractor in Fremont. There was "nothing spectacular or unusual in its performance," a reporter from *Gas Power* noted, though it did have "some features other manufacturers can profitably study." But it was also recognized that most farmers "would have taken the machine without question, just because it was built by Henry Ford."

Ford's publicity, coming free from newspapers and trade journals, an International Harvester historian wrote thirty years later, "was so at variance with the actual facts that those who were really posted on the situation was dumfounded [*sic*] at the stupidity which was evident."[18]

But even the critical report from *Gas Power* understood the benefits of Ford's tractor. "The icing seemed to be quite general among manufacturers that Mr. Ford's tractor will do more to eliminate the fly-by-night concerns and their products than any other factor. The Ford tractor will be a means of stimulating the buying of all makes of tractors."[19]

Many farmers, not quite sure *what* they were waiting for, simply waited.

Deere president William Butterworth was not in Fremont and was unable to make the September 1916 directors meeting in Moline. He was on the road, meeting with some of the company's bankers. Butterworth was well aware of the potential impact of a tractor from Henry Ford and asked Burton Peek, the company's general counsel, to stop any further discussion of the tractor in his absence. "There is one thing I want to ask you to look out for," he wrote, "and that is any further action by the directors relating to the building of a tractor. I am opposed to any steps being taken looking toward the manufacture by any of our factories of a tractor. I have acquiesced in the experimental work which has been done, but I am beginning to feel that we are wasting the stockholder's [*sic*] money in going any further with it."[20]

Butterworth did not address the strategy ardently promoted by George Mixter, the man with the most intimate knowledge of Deere's production capabilities. If Deere did not build a tractor, was there another way?

8

"A WAR TO END ALL WARS"

Shareholders at the Ford Motor Company would have agreed with William Butterworth's assessment on the speculative and unprofitable state of the tractor industry. Ford's board of directors continued to challenge its founder and president on mounting tractor expenditures.

In 1915, Ford and James Couzens, a trusted ally connected to the Ford Motor Company since its inception, inaugurated a series of secretive land purchases on tracts near the Rouge River, southwest of Dearborn. In all, they secured nearly two thousand acres from dozens of farmers and other landholders. When information about the "elaborate plan" began to leak, Ford misdirected the media, sharing that the new factory was being constructed to build tractors and furnaces to convert ore to iron. When completed, it would employ twenty thousand people, be less than one-third the size of Highland Park, and have a payroll of $30 million. Housing developments around the proposed site exploded. The Grand Factory subdivision sold lots for $500, including water, sewer, gas, and electricity, all for only $50 down and $10 per month.[1]

Ford's plans were also much more elaborate and, according to his shareholders, unnecessary. John and Horace Dodge owned two thousand shares of the Ford Motor Company, nearly a quarter of the shares held outside of the Ford family, and looked to protect their initial $10,000 investment, which to date had earned them tens of millions of dollars in dividends. The brothers had taken early proceeds and had begun their own competitive line of automobiles in 1913, their facilities built and still funded almost exclusively from the dividends paid by the Ford Motor Company. A pending change in terms would reduce their returns by 90 percent. Net profits at the Ford Motor Company approached $60 million in the year ending July 1916 alone.

The debate came to a head in November, the day after the wedding of Ford's son, Edsel, to Eleonore Clay on November 1, 1916. At both the wedding and the evening reception, Henry and John and Horace Dodge seemed to put their growing ideological differences aside. The following day, Ford was notified of a filing in the State Circuit Court by the Dodge brothers against the Ford Motor Company, specifically his efforts to eliminate dividends and redirect funds into a new factory, which, as the Dodge brothers were led to believe, would build Ford's farm tractor.

Ford always distrusted investors: they stood in the way of reinvestment by collecting dividends while not materially participating in the company's operation, most notably Ford's most ambitious gamble to date, the erection of a new factory. Shareholders knew that his intention to build tractors at the new location was more than misdirection. "When I get this tractor going there won't be any men left to work in munition factories," Ford said assuredly. "I'll have 'em all making tractors."[2]

John Dodge, eldest of the brothers, called on Ford to "discuss his tractor plant," gaining assurances that no Ford Motor Company resources would be used in its manufacture or sale, but also getting confirmation that he was indeed prepared to eliminate dividends. "I told him," John Dodge later testified, "that if he proposed to carry things to such an extreme as that he should buy out other stockholders; then he could run the business as he saw fit." The brothers alone stood to earn $35 million from such an arrangement, Ford testifying that they would "harass me in anything I tried to do" if he did not pay their price.

Ford and his attorneys would be wrapped up in the case for nearly a year, during which time Ford was restrained from using any of the Ford Motor Company's cash resources toward factory expansions. That included tractor production, a judgment now with greater implications for the people of Great Britain than for farmers in the United States.

Though separated by an ocean and thousands of miles, Americans followed news of the Great War from their homes and homesteads. In April 1916, President Wilson called on Germany to end surprise submarine attacks, some on non-war vessels that included American ships. That summer, along a fifteen-mile line in northern France, the British suffered more than sixty thousand casualties in a single day of fighting. The Battle of the Somme turned into a five-month stalemate, the British and French armies alone suffering more than 500,000 casualties by the time it ended. The offensive ultimately advanced the British and French lines five miles.

In the United States, "Ford-osis" had a full grip on the nation, and Ford's long-promised tractor business, not to mention the business of farming, was set to benefit. "It's on the brain and in the blood of the American people," offered the *Detroit Saturday Night*, which coined the phrase. "They gobble the Ford stuff, and never stop to reason whether they like it, or whether it has any real merit in it . . ."[3]

Former president Theodore Roosevelt complained that Ford received more attention than President Wilson and was a distraction to the country. It did not come naturally at first, but Ford had learned how to expertly curate the attention, cultivating media coverage around his personal ability to deliver inexpensive, quality transportation. Progressive decisions like the institution of the $5 workday, an unprecedented move, further endeared him to the working class. The transformation of Ford the folk hero, the man whose first contemporary biographer credited as "inventor, creator, manufacturer, humanitarian, and public servant" and who will "live in history . . . as one of the Makers of America," was underway.[4]

Ford's isolationist views in support of President Wilson in the opening years of the war had not endeared him to the British people. When the United States approved a £100 million loan to the British government, Ford threatened to withdraw funds from all the British banks that participated. "If I had my way," he was quoted in the English press late in the summer of 1915, "I would tie a tin can on the joint Anglo-French Commission and chase it back to Europe. The best thing would be for all of Europe to go bankrupt; then the fighting would stop."[5]

Soon after, he embarked on what came to be called the Peace Ship, a publicity tour organized at the urging of peace advocates from across the country. "I will do everything in my power to prevent murderous, wasteful war in America and the whole world," he told the *Detroit Free Press*, personally convinced of his ability to do it. When President Wilson, "a small man," Ford afterward said, refused to legitimize the effort, Ford chartered a ship and financed the mission himself.[6]

Populist American newspapers predicted his $1 million "educational campaign for peace and not war" to be his "big work for the rest of his life."[7]

Announcement of the Peace Ship was both celebrated and dismissed domestically. The list of promised envoys included inventor Thomas Edison, investigative journalist Ida Tarbell, social activist Jane Adams, three-time Democratic presidential candidate Williams Jennings Bryan, and more. Unfortunately, none of them actually went. "The fact that a man can make a cheap automobile is not necessarily a qualification for becoming a world leader," commented the Philadelphia *Record*. Former New York Senator Chauncey Depew dismissed the charade as a simple publicity stunt.[8]

Despite the criticisms, the Peace Ship left in October 1915 with a list of passengers representing nearly every type of social reform effort in the country. Ford held twice-daily press briefings during the thirteen-day trip across the Atlantic. He arrived to surprisingly little fanfare.

Ford used a December 22 briefing to pontificate on the importance of tractor manufacturing.[9]

A day after the ship's arrival, Ford himself quietly returned to the United States. He would soon find a more receptive audience for his plan to put a tractor on British soil.

England's Ministry of Munitions, M.O.M., was created by the Munitions of War Act in 1915, tasked with the coordination of manufacturing and distribution of munitions, a reactive response to the crippling shortage of artillery shells and ammunition. The department was put under the liberal direction of David Lloyd George, most recently Chancellor of the Exchequer, soon the prime minister of Great Britain. He was a man who got things done, now tasked with feeding the war machine, which he did by putting women to work in factories, regulating private industry, and mobilizing trade unions, all in a successful effort to increase the rate of production.

In 1915, the *London Times* estimated that eighty thousand farm hands had already left for military service, and tractors were in high demand.[10]

"The situation was a grave one," Ford later wrote.[11]

Thousands of farm horses were being pressed into service to provide logistical support. Trained operators of steam engines were one of the many classes of labor shortages. A series of hurricanes and tropical storms in 1916 added to Britain's food scarcity.

Ford's plan for British food production was built on a concept first proposed after a European vacation he took in 1912. The plan at the time called for the erection of Ford's first automobile factory outside of the United States. The location was Cork, Ireland, chosen not because of its strategic location, though it had its advantages, but because it was the ancestral home of the Ford family. Cork rivaled Belfast and Dublin as one of the great industrial cities of Ireland, the former with a population of more than 75,000 people at the turn of the century. Visitors traveled to Cork via train, along the river Lee, a thirty-minute ride from the port city of Queenstown. The already famed Blarney Castle was less than ten miles away.

Ford's trusted representative in Great Britain, Percival Perry, preferred a location somewhere in Southampton instead, but the war changed that.

Lord Percival L. Perry (he would be knighted in 1918) was a decade-long Ford partner in England. Born in Bristol, in western England, he moved to London at the age of seventeen and joined the Central Motor Company, a distributor of Ford cars, among other things. He became managing director in 1906 and, having trouble financing the purchase of Ford automobiles, traveled to Detroit to meet with Ford himself. The meeting started a long relationship between the two entrepreneurs. Ford called on Perry a few years later to establish a Ford branch in Britain, and in 1914 they opened a factory in Trafford Park, Manchester. During the war, Perry served as Ford's eyes and ears.[12]

In 1916, newspaper magnate Lord Northcliffe, leading a British War Mission to the United States, visited Ford's small, hastily assembled tractor factory in Dearborn, one of many visits made by the British group while on American soil. Lord Northcliffe, the most powerful publisher on British soil, easily succumbed to the sheer scale of Ford's capabilities. Despite Ford not yet having a tractor to sell, or a suitable facility to build it, Northcliffe saw its beginnings in the work already accomplished by Galamb, Sorensen, Farkas, and others. Ford's ability to quickly scale and increase production were never in question.

Approvals for construction of a tractor factory in Cork came in the spring of 1917. A month later, after the sinking of American merchant ships by Germany, President Woodrow Wilson asked Congress to join Britain, France, and Russia and declare a "war to end all wars."

Perry wired Edsel Ford the following day, urging him to send Charles Sorensen and all their tractor drawings. ". . . The matter is very urgent," he telegrammed, ". . . National necessity entirely dependent Mr. Ford's decision."[13]

The sentiment was far from universal. The new minister of the Munitions Board, Lord Milner, thought the Ford alliance suspect from the beginning and urged the importation of tractors from a multitude of American companies that, most importantly, already had inventory and could ship promptly.

"The development of agricultural motors is a matter of the highest importance to the whole of the British Empire at a time when the cost of labour is going up by leaps and bounds, and a competitive shortage both of men and horses appears inevitable," Milner noted.

American tractor brands already had a foothold in the United Kingdom, tested at agricultural expositions and farm demonstrations just as they had been in the United States. To date, they were imported in small numbers, sometimes carrying their American names and sometimes rebranded. The Highland and Agricultural Society of Scotland demonstration in 1915 in Stirling, a town famous for William Wallace's defense against the British in 1297, featured three "internal combustion tractors of American origin," one British steam tractor, and one British motor plow. That lone British motor plow, the Wyles, "received on the whole the highest commendation of them all," reported the *London Standard.* Most British firms were absent because their shops had been converted to produce wartime munitions.

International Harvester's Mogul tractors first appeared in a 1913 demonstration, the "subject of most favourable comment." It cost twice as much as the domestic Wyles, which sold for £170, but plowed three furrows instead of two

Mogul tractor supplied by the American Red Cross to convalescents in Contrexeville, France. Library of Congress, 2017682010.

and could drive a full-size threshing machine. Harvester's Titan line arrived in 1917.

The Waterloo Boy Model R tractor, built in Waterloo, Iowa, a descendant of the first successful gasoline tractor built by John Froelich in 1892, arrived in Northern Ireland in 1917. There it was repainted olive green, set off with deep red spokes, a gray engine case and fuel tank, and rebranded as the Overtime tractor by L.J. Martin of the Overtime Tractor Company, London. The tractors burned low-cost paraffin, similar to American kerosene, which was preferred to petrol due to its lower cost. An estimated three thousand Overtime tractors were in use in Great Britain by the end of the year.[14]

Henry Ford & Son shipped and assembled a single tractor in April 1917, just in time to join a field of fourteen other American and seven French tractors at the second day of a two-day Paris field trial with the French Minister of Agriculture in attendance. A second Ford tractor sat in a French port. The Ford's white paint glistened as a canvas for the message, "Prosperity, Industry," which had been painted across the hood. The tractor "attracted a lot of attention and worked with great regularity." The Paris tractors were for demonstration purposes only.[15]

On May 3, Sorensen, five engineers, and train cars full of unassembled tractors, parts, and implements, left the United States aboard the *Justicia*, a British troopship that would be torpedoed and sunk a year later. They were scheduled for assembly by the first of July 1917 in Ford's automobile factory in Manchester, though not in time for three tractor demonstrations hosted by the Highland and Agricultural Society at Perth, Edinburgh, and Glasgow. Three-fourths of the twenty-nine tractors in the field were American, including the Samson, Mogul, Titan, Parrett, Overtime, Bull, the Big Four 30, the Moline Universal, and the Wallis Junior Cub. A local paper considered the Wallis to be "undoubtedly, from an engineer's point of view, the best example of tractor construction present." Fifteen of the entries ran on four wheels, six on three wheels, and four on tracks. Eleven of the machines required two men to operate. No effort was made to judge the machines individually or rank them, but only to determine, as a whole, the "important respects in which farm tractors and plows may be made useful in tillage operations."

The committee instead outlined a series of recommendations based on the whole of the trials. Price should not exceed £300, or $1,460. Horsepower should exceed twenty at the drawbar, and wheeled tractors were preferable over tracks. Enclosed gears were preferable, and speeds of two-and-one-half, and four miles per hour forward, with reverse, "appear to be most generally useful."

That summer, a deal also seemed to be in place by the Food Production Department to purchase at least 500 Bull tractors, "one of the largest and most successful American agrimotors . . . that is, in fact, years ahead of certain other tractors that are strongly favoured . . . ," reported *The Commercial Motor*. The paper considered the Bull a high-performing machine and, more importantly, an available machine.

"Why should English farmers have been deprived of Whiting-Bulls [the distributor], which are known and found to be all right, while Ford gets ready? Who is the real obstructionist?"

Sir Richard Winfrey, Parliamentary Secretary to the Board of Agriculture, responded in line with American reports on the Bull, critically noting a "lack of power and its unfortunate design; both these tend to make it impossible to work steadily or satisfactorily on anything but the most level and moderately light ground."[16]

Nothing could tarnish Ford's reputation among the people of the United States, certainly not delays in tractor shipments to Great Britain. The American press continued to mythologize the flivver king (a reference to the inexpensive Model T), pitting Ford against Krupp, Germany's primary manufacturer of armaments, and the German submarine against the American tractor. The "submarine is the engine of starvation; the tractor is the engine of plenty," wrote B.E. Ling in an editorial in West Virginia. "The submarine stands for destruction; the tractor for production. It is the submarine's mission to starve the allied world into submission to Germany; it is the tractor's purpose to nourish it for victory over kaiserism."

"Henry Ford has invented the tractor. He has perfected it."

Ford did not dispute the claims, and through the continued and growing success of the Model T and occasional reminders of his tractor's imminent availability, continued to convince patriotic American farmers to wait a little longer. "My tractor is a proved success," confirmed Ford, despite having built only fifty to date, and with no contract yet in place. "I will accomplish all I have sought for it to accomplish."

The tractor, said Ford, was not a "weapon of war, but a blessing of peace."

"Ford will triumph over Krupp," concluded Ling.[17]

In late May 1917, with a small shipment of unassembled Ford tractors en route to Great Britain, twenty-two German Gothas, armed with thirteen bombs each, flew noisily over London, destroying homes, churches, and businesses, killing at least fifty-seven people and inflicting new horrors on the British people. This latest attack quickly diverted government attention from a fleet of land-based tractors to an urgent need to reinforce the air fleet.

Recently approved plans to build Ford tractors in Ireland came under renewed scrutiny, and Ford stalled on the delivery of his tractor. Unlike his Model T, which he considered the perfect automobile, Ford did not consider the tractor to be yet perfected. Lord Milner desperately requested a new arrangement, suggesting they be built in Dearborn and shipped fully assembled. Sorensen, on behalf of Ford, agreed, under the condition that Britain would pay to ship them. Sorensen promised fifty assembled tractors within ninety days, plus five thousand more at a price of $50 over the cost of production.

Lord Northcliffe implored Ford, "We understand your objection," pointing to military weapons being rushed to production despite imperfections. "We need a tractor. Yours is the best we can get. We can't wait for the perfect tractor; we need what's available and we want you to produce it."[18]

Ford acquiesced.

"It is generally known that our tractor . . . was put into production about a year before we had intended," Ford later rationalized. In his version of the story, Great Britain, on the verge of starvation from the ravages of war, came begging for help to feed her people.[19]

An "emergency extension to our plant at Dearborn, outfitted with machinery that was ordered by telegraph and mostly came by express, and in less than sixty days the first tractors were on the docks in New York in the hands of the British authorities," remembered Ford.

American farmers continued to wait.[20]

9

"FIRST CLASS ALL THE WAY"

Introduction of the inexpensive and lightweight Bull tractor, the long-awaited appearance of Henry Ford at Fremont, and incremental growth in tractor sales nationally encouraged competition and drove increased production. New models from the Wallis Tractor Company, the Huber Co., the Eagle Manufacturing Company, the Parrett Tractor Company of Chicago, and the Waterloo Gasoline Engine Company were just some of the debuts in 1916. Each worked to secure a small part of the fragmented market.

Deere president William Butterworth, humorous, plainspoken, and straightforward, would tell anyone who would listen that "you can't sell a plow to a hatter." He knew too well that farmers were knowledgeable and could identify an inferior product when they saw one. But tractors challenged their judgment like no piece of farm equipment had before. They were complex, prone to mechanical failure, and constructed of non-standardized, often crudely made parts. And it was likely the most expensive purchase a farmer would ever make.

There were multiple tractor classes, typically denoted by two numbers representing drawbar and belt power, and a third categorization based on the number of standard plow bottoms it could pull from the drawbar. Soil type, plowing depth, and other factors impacted actual use, but advertising typically provided an accurate accounting. Smaller tractors, like the Bull or International Harvester's Mogul 8-16, were rated capable of pulling a two-bottom plow. Ford's slightly more powerful tractor fell into that class as well. As development of Joseph Dain's tractor further convinced Deere's designers of the necessity of a rugged, reliable tractor, focus on the three-plow market, which would be less crowded after Ford's arrival, looked more appealing, as long as the issue of price could be resolved. So far engineers, accountants, and sales managers had been unable to reconcile expectations with reality.

One thing was clear. "The farmer is gradually beginning to realize that the tractor is coming to stay," noted the *Farm Implement News*, "that it is not some miracle machine that will take all the hard work out of farming, but rather a machine that places farming upon a better business basis and a machine regarding which he must be well informed."[1]

A movement toward regulation and standardization of both advertising and field performance began to emerge from the 1916 demonstration in Fremont, an effort to eliminate manufacturers that over-promised the capabilities of their products. Not surprisingly, Ford was in the middle of the debate, though in this case as the victim of speculators capitalizing on the Ford name.

The formation of Henry Ford & Son, a company created independent of the Ford Motor Company in 1915, coupled with the continued promise of a farm tractor not yet delivered by Henry Ford, was an invitation for a fraud that began almost immediately. The instigator was W. Baer Ewing, who created the Ford Tractor Company in Minneapolis, naming the company after Paul B. Ford, a local electrician, husband, and father of two who could not believe his good fortune. In reality, Ewing was dodging creditors when he set up his new enterprise, selling stock for the "Famous Ford Tractors" to an expectant public waiting for Henry Ford's tractor. It was "The Machine the World has been waiting for."

Its first tractor, the Ford Model B, was sold in July 1915, and a year and a half later the company claimed tractors "in successful operation" in thirty-seven states and was in the process of raising $10 million through a public stock offering. The Ford Model B sold for $495 and in time would have long-lasting implications for the industry both as the result of the company's deceptions and the tractor's failures.

Ewing and his sales manager, C.M. DeVeau, even demonstrated their new Ford 8-16 tractor at the National Demonstration in Fremont in August 1916, though whether Henry Ford or his contingent was aware of the fraud under their noses was unreported.[2]

That year, Wilmot Crozier, a former educator, returned to the family farm in Nebraska and purchased a Ford tractor for $350. He was thrilled at the performance of his Ford Model T automobile, and a Ford tractor was an obvious replacement for his horses. But what he bought was not the Ford tractor promised by *the* Henry Ford, but an 8-16 two-plow tractor from the Ford Tractor Company of Minneapolis. It broke down.

Crozier demanded a replacement, which fared no better, then purchased a secondhand Bull tractor. When it failed to live up to the advertised specifications, he bought a Rumely Oil Pull, a three-plow tractor, which, he said, could have pulled five plows. It actually exceeded expectations.

Crozier later shared that he had "followed many a queer-looking contraption around the demonstration fields that purported to be able to replace my long-eared mule in front of a gang plow." He had grown convinced, with each passing year, that the tractor was "the agricultural implement the American farmer had been looking for, lo these many years." After several investments in underperforming machines and then "in a machine that would really do what the company said it would . . . I began wondering if there wasn't some way to induce all tractor companies to tell the truth."

Crozier's crusade would forever change the tractor industry and over the next three years provide it much-needed credibility. He was elected to the first of three non-consecutive terms in the state legislature beginning in 1919 and began work on legislation with State Senator Charles J. Warner of Waverly. They pulled in Professor L.W. Chase, a judge at the Winnipeg Motor Competitions and then the first National Tractor Demonstration in Nebraska, and now a major in the United States Army, to help write the bill. The Tractor Test Bill passed in 1919.

According to the bill, "A stock tractor of each model sold in the state be tested and approved by a board of three engineers under State University management." Secondly, "Each company, dealer, or individual offering a tractor for sale in Nebraska shall have a permit." And, third, "A service station with a full supply of replacement parts for each model tractor shall be maintained."

As a direct result of the legislation, the Nebraska Tractor Test Laboratory was established at the University of Nebraska in Lincoln. In 1919, the American Society of Agricultural Engineers and the Society of Automotive Engineers created standards for belt speeds and drawbar hitch heights, and the Nebraska tests supplemented standards through published test results.[3]

In the meantime, development and competition continued to accelerate at an unprecedented pace.

William Butterworth's resistance to tractor development at Deere was driven by the continued sheepishness of the banking community. The company's quarterly dividend had fallen from an average of 20 percent a decade before to scarcely more than 1 percent in 1915. New profit-sharing initiatives had just begun, including an employee stock purchase plan. An employee health benefit plan was introduced in 1908. All required capital.

Still, with an increasing sense of urgency, experimental work continued. J.S. Molstad and George Schutz from Deere's Minneapolis sales branch gave an encouraging report on a multi-month field study of Joseph Dain's tractors in Aberdeen, Fargo, and Minot. "We are firmly of the opinion, providing the clutch collar and bevel pinion sleeve trouble can be eliminated, and not taking into consideration the power plant or the question of belt power, that we have in the Dain very much the best tractor on the market." They offered five reasons:

1. Its light weight
2. Using four chains to drive the wheels puts less strain on the sprockets and wheels, making it two to three more times durable than sprockets and easier for farmers to replace
3. Being able to change speed without "clashing" the gears

4. Being able to change speed without stopping, saving gas and trouble
5. Its foolproof simplicity and accessibility

The "matter of price should be forgotten for the present," they recommended, a controversial recommendation after four years of designing toward a specific price point. "Go ahead and build the tractor—first class all the way through, using extra good magnetos, carburetors, etc., as well as making it extra good in other details, and when that is done if the price must of necessity be $1,500 to market them profitably, let's sell them for that."

Before the board of directors, Dain flipped through a series of charts comparing his tractor to competitive machines tested during the trials. In Aberdeen they saw demonstrations of a six-plow Gray tractor, a Waterloo Boy, and a Heider, which after nearly four hundred acres of work, was "about ready for the scrap heap."

A Mr. Harrington of the Fargo Implement Co. had sold seventeen Waterloo Boy tractors—seven in 1915 and ten in 1916, and thought he could sell twice as many if he could find enough knowledgeable mechanics, but they were scarce and expensive. Besides, he shared, the risk was too great. If the tractors had mechanical trouble, customers would blame him, not the manufacturer. And "they all seem to have an undue amount," he confided, though recognizing also that if he "had not sold these tractors he probably would have sold only four or five of our Pony Plows instead of the seventeen that he sold this year."

"The Waterloo Boy model R tractor," Schutz and others concluded, could not pull more than one sixteen-inch sulky plow and "stand up." The Dain performed the best of the three, with the operator recommending a stronger, heavy-duty motor. Molstadt and Schutz concluded that "our dealers need a tractor badly, but they must have a good one."[4]

Again, Deere's board, despite Butterworth's reluctance, approved further work on the Dain tractor.

A new machine form had garnered attention in recent years, and now held the attention of designers at both International Harvester and John Deere. The motor cultivator, as it was commonly called, was a bridge between the horse and the tractor. In fact, several manufacturers were mounting engines to existing cultivators, an implement similar in design to a plow, but with pointed shovels that turned over weeds and buried them in the soil between planted rows.

The most promising motor cultivator was the Universal, an original design of an Ohio farmer named Truman Funk, who formed his company in 1913. That year, he built fifty motor cultivators, followed by one hundred the following year. International Harvester declined an opportunity to acquire the line, which eventually was sold to the Moline Plow Company. Its headquarters sat across the street from John Deere's headquarters in Moline.

A newly redesigned and updated Universal, a "unique type of tractor," debuted at the National Farm Demonstration in Fremont in 1916, just a few months after production began in a new factory in Rock Island, Illinois. Advertising broke it down. "As powerful as five horses; as enduring as seven horses; costs less than four horses; requires less care than one horse; less room than one horse and eats only when it works."

The new two-cylinder model was rated at twelve-belt horsepower and was available with a variety of implements—a two-row cultivator, two-bottom plow, disc/harrows, grain drills, a corn planter, and a ten-foot grain binder. The Universal was considered an entry-level machine familiar to horse farmers but with a number of concepts far ahead of its time, including electric lights and a starter.

The Moline Plow Company sold enough machines that at times the company could rightfully claim to operate the world's largest tractor factory—it was a technicality as International Harvester's tractors were coming from several different factories.[5]

Meanwhile, Deere's tractor development continued to reflect the still-fragmented state of the industry, and proponents of the motor cultivator, led by Theo Brown and engineer Walter Silver, secured funding for development. Deere had been designing cultivators for sixty years, and the concept offered as much a bridge for their tractor concepts as it did for horse farmers looking for a motorized alternative.

The Moline Plow Company's Universal helped inspire the general-purpose tractor. This one is working in corn in 1918. Wisconsin Historical Society, WHS-24694.

In the midst of the Civil War, Deere had acquired the manufacturing and distribution rights for the Hawkeye Riding Cultivator, winning a first premium at the Iowa State Fair in 1863. Improved versions over the next few years brought first premiums at state fairs in Iowa, Illinois, and Wisconsin. Pulled by two horses, it was practical for use even in high corn, as tall as five or six feet, without damaging the crop. The seat was "sufficiently high and convenient" so operators' legs did not brush against the corn, and provided full visibility to the front-mounted cultivators.[6]

Brown, Silver, and even Joseph Dain all developed drawings for motorized cultivators at Deere, with a variety of one-, two-, and three-row versions tested as early as 1916, and continuing off and on again through 1921. None made it to production.

Brown's cultivator incorporated a New Way air-cooled engine from Lansing, Michigan, which proved deficient. Next, he tried an engine designed by the Associated Manufacturers Co. of Waterloo, Iowa, but that too was inadequate.

Finally, Brown went to Walter McVicker, the brilliant young engine designer already working on a new, more powerful tractor engine for Joseph Dain's tractor. In January 1917, the board approved material to build fifty motor cultivators with the new McVicker engine, directing that "twenty-five machines be built as rapidly as possible."

Privately, while preparing two machines for field tests in San Antonio, Texas, Brown lamented that "the tractor is the chief source of all our troubles." Poor weather brought ideal test conditions to East Moline in early March, and the two Tractivators, as they were now calling the prototypes, dragged stoneboats, stone weights used to simulate drag, through a muddy field behind the Marseilles Works. Tests were promising, but production challenges began to pile up. The Associated Manufacturers Company did not have the capacity to fill the engine orders from McVicker, and local wheel manufacturer French & Hecht was "having trouble rolling tires for the tractivator."

Theo Brown's Tractivator in a July 1916 field trial. From left to right are many of Deere's early tractor development leaders: Dr. Warren E. Taylor, Theo Brown, H.H. Bliss (Deere's patent attorney in Washington), Gus Bischoff, George Mixter, Floyd Todd, Joseph Dain, Sr., and Harold Dineen. Courtesy John Deere.

Later that spring, Dr. Warren E. Taylor, head of the Soil Culture department, escorted a reporter for the *Eastern Dealer*, a popular farm journal, around Deere's many Moline and East Moline area experimental farms. On the "corn farm," his "horseless farm" as it was described, the Tractivator, "which pulls a plow and shoves a cultivator and reaper, was working fine." Taylor predicted that "some day soon this will be on the market, and it will be a dandy when they get it ready."[7]

By mid-April, Deere had sourced enough engines to build and ship twenty-five more Tractivators to six separate sales branches: Moline, Minneapolis, Indianapolis, Omaha, Kansas City, and St. Louis—all to be used to cultivate corn in the spring and the summer and then to cut hay in the fall. If all went according to plan, the Tractivator could sell for as little as $475. George Mixter's only concern was that they might not be able to build them fast enough for the following year.[8]

Brown proudly took Joseph Dain to watch the Tractivator operate with a mower in April, and soon customers would get their first glimpse in the pages of an educational book authored by Taylor and published annually by John Deere, *Soil Culture and Farm Methods*. Farmers could secure the book by sending in a free coupon provided by their local dealer and a fifty-cent stamp for delivery.

Compiled from available national data from universities and leading experimental farms, the educational resource was part of the growing literature intended to educate farmers on soil, "the source of all wealth," and implement types and operation. It was another outcome of the many experimental farms testing the latest in farm technology, all featuring John Deere equipment, of course.

For all the problems of growing, cultivating, harvesting, and transporting crops, there "seems to be but one solution to the problem," the 1917 edition of the book offered: "Production must keep pace with the increase in our population if the nation is to survive and prosper . . . The responsibility rests with the farmer."

A section on "The Farm Tractor," noting there being "135 different makes on the market, and more in a state of development," included a John Deere tractor for the first time. "The John Deere 'Tractivator,' a light tractor which promises to fill a gap and absolutely revolutionize the motive power of the farm," was "as easy to handle as a Ford car and can be operated at a minimum cost."

But before the books shipped, there was a revision. When results of the twenty-five Tractivators returned from the sales branches, tests revealed that even at $475 retail, the machine offered customers little to no savings over the operational costs of a cultivator and two horses.

On page 41, an "IMPORTANT NOTICE" in bold red letters was pasted below a line drawing of the Tractivator. "The great demand for materials of all kinds, due to the war, makes it unwise to enter new fields of manufacture at this time," it read. "Therefore, the offering of the Tractivator to the public has been postponed for the present."[9]

Deere's first tractor entry was again stalled, though after four years, developmental work on the Dain tractor was drawing to an end. The revised McVicker engine "now looks alright," Joseph Dain told the board, but "as to the cost of the tractor as a commercial possibility, even he had some doubt." The price tag, now at least $1,200 and growing, looked out of reach, especially compared to the Tractivator. Regardless, he "did not believe Deere & Company could afford to drop it and not have a well-developed tractor up their sleeve."[10]

In June 1917, Dain made his best pitch on the all-wheel-drive tractor, telling his colleagues that the "day of the cheap tractor is about over. The work the tractor must do precludes the use of cheap materials and workmanship." His tractor weighed less than competitors' models and had more power at the same speeds. Said Dain: "The most important difference is that they [other manufacturers] do not use the same class of workmanship or materials as we must use, as must everyone else to have their tractor a success in the hands of the farmer."[11]

By September, there were more prototypes in the field in Minneapolis, North Dakota, and South Dakota. They considered the tractor in Minot to be the "best on the market." It had all-wheel, four-chain drive, which was much quieter and more durable than gears. It could change speed without stopping, even under load, thereby saving time, and it had foolproof simplicity and accessibility.

Joseph Dain's all-wheel-drive tractor with a John Deere grain binder, circa 1916. Courtesy John Deere.

One of the tractors shipped to Huron, South Dakota, and was tested on the farm of John Deere dealer F.R. Brumwell. It plowed 110 acres, harvested 260 acres, and pulled five wagonloads of stone about twelve miles to Huron on two different occasions. The plow used was a John Deere No. 5 Pony tractor plow with model NA 14-inch bottoms working seven inches deep. Even when the plow was set down to eight, and even ten, inches deep, the tractor pulled it with no engine knock or undue wheel slippage. Brumwell, a farm implement dealer and proprietor of a local sawmill, was enthusiastic about this tractor and believed it was much better than any other he had seen. He bought three tractors with the new motor: one for his farm and two for customers.

The first tractors with the Waukesha engine were satisfactory in every way except that they lacked power. When they were changed over to the McVicker-designed engines, the tractors proved to do all that was required of them. The final drive chains had also been strengthened to handle the higher horsepower. Of the four 1917 tractors with the new engine, three went to Huron and one to Minot. Field tests convinced the board to order "not over 100" of the Dain tractors to be built by an outside firm. Soon after, this was altered in deference to in-house assembly in East Moline.

Ironically, now there was concern that the three-plow Dain would be too small and that "we would probably have to eventually make a bigger one."

Dain told the board in September that he had been working on a supplier for "having probably one hundred tractors built outside, but had been unable as yet to make much progress." After a strong endorsement by C.C. Webber, manager of the Minneapolis sales branch, and Floyd Todd, manager of the John Deere Plow Works, the board moved that "Mr. Dain be authorized to continue his negotiations with the thought in mind of buying not over one hundred of the Dain Tractors outside with the view of continuing the development of the tractor line, the purchase to be made on a fixed basis."[12]

Industry sales reached nearly 63,000 tractors that year, double 1916, built by 124 companies. The motion to build Joseph Dain's all-wheel-drive tractor passed unanimously, but not without uncertainty for the future.

10

"ENGLAND GETS THEM FIRST"

By 1917, world events, farm labor shortages, and food conservation added increasing urgency for advances in the business of food production. With American entry into the war, young men and women were leaving the family farm in even greater numbers, now for the front lines and with the possibility of never returning. Nearly five million would serve before the war's end. Still, crops had to be planted, cultivated, harvested, and delivered. Enough tractors were now in operation that customers who were ready to buy no longer had to wait for a demonstration but instead could narrow down their prospects from trade publications, the experiences of neighbors, and recommendations from trusted local dealers.

On the home front, the newly created United States Food Administration encouraged meatless Mondays, reduced consumption of wheat and sugar, and greater consumption of potatoes, which were heavy and more expensive to ship to American armies overseas. Posters were plastered on walls in post offices and government buildings urging even "little Americans" to "do your bit" by eating corn cereals and "save the wheat for our soldiers."[1]

Addressing a growing list of priorities, the National Implement and Vehicle Association met eleven times over the course of 1917, mostly in Chicago. Association leadership included representatives from International Harvester, John Deere, J.I. Case, Avery, Emerson-Brantingham, and other leading farm equipment makers. Deere's Joseph Dain, the association's president, ran the meetings.

Beginning January 31, a handful of members attended the three-day United States Chamber of Commerce Annual Conference in Washington, DC, at the Willard Hotel, a meeting place for the city's political writers and beat reporters since the Civil War. President Wilson could not attend. Instead, former president and future Supreme Court justice William Howard Taft spoke on the "League to Enforce Peace."

At the May meeting, a special delegation met with editors of the leading trade papers to "bring about their co-operation in prosecuting a propaganda to education along food supply lines." The Secretary of Agriculture was the featured speaker.

That summer, the organization absorbed the smaller National Association of Tractor and Thresher Manufacturers.

At the September 13 meeting, sixteen companies were approved for membership. They were nearly all tractor manufacturers, including the Wolverine Car and Tractor Co. of Detroit, the Gray Tractor Company of Minneapolis, and the Allis-Chalmers Manufacturing Company of Milwaukee. Membership for the Ford Tractor Company of Minneapolis, still deceptively selling under the Ford name, was declined, as was the Square Turn Tractor Company of Norfolk, Nebraska, the latter pending "further time for investigation." The Wolverine Car and Tractor Co. was later dropped, as were many more, for failure to pay dues.[2]

Henry Ford & Son had not applied for membership.

On October 6, 1917, with little fanfare, eighteen men, among them Henry Ford, Eugene Farkas, and Charles Sorensen, posed with the first fully assembled Ford tractor destined for England. The tractor rested on a wooden platform, covered by a fitted canvas with the stenciled words "Ministry of Munitions, Agricultural Tractor Factory, Trafford Park, Manchester, England." The tractor itself had no markings to denote its source. It came to be known simply as the M.O.M. (Ministry of Munitions) tractor.

Patient American farmers, a growing number of whom enjoyed the new-found opportunities and freedoms facilitated by their Model T automobiles, continued to wait.

"England gets them first," supportively printed a West Virginia newspaper, unaware of the small volumes and shipping bottlenecks endured by both sides. "England gets them first because Germany is trying to starve England first."[3]

By December 1917, there were 1,660 government-owned tractors in use in the United Kingdom, an increase of only 220 machines since October. Only

The first Ford tractor for wartime Great Britain, 1917. First row front (men seated on planking, far right: Joseph Galamb. Second row (men standing on ground), left to right: unknown, Ernst Kanzler, Henry Ford, Charles Sorensen, Eugene Farkas. Image from the Collections of The Henry Ford.

fifteen thousand acres had been plowed. More American tractors were in private hands, purchased directly through distributors.[4]

Ford later claimed the entire shipment of tractors arrived within three months, but in actuality, only 254 machines were delivered by the end of 1917, and only 3,600 by March 1918.

In February, Henry Ford cabled Lord Northcliffe to complain of shipping delays, threatening to sell the three thousand tractors sitting at docks in Baltimore and Philadelphia to customers in the United States or Canada. It was not until near the war's end that all seven thousand, now including an order from Canada for one thousand tractors, were delivered.

The long-anticipated Ford factory under construction in Cork, Ireland, was again delayed and did not begin operations until the following summer. For now, tractor assembly was canceled, and Ireland would build only the world-famous Model T.[5]

* * *

With his commitment to Great Britain more or less fulfilled, Henry Ford was making final preparations to sell his tractor in the United States. Ford's tractor was originally conceived from the mechanical principles of the automobile, with intention to share parts, engineering concepts, and even an engine, but its designers evolved the design as they continued to understand the unique demands of the farm. "As in the automobile, we wanted power—not weight," wrote Ford. "The weight idea was firmly fixed in the minds of tractor makers." The automobile was built for carrying people. Tractors were built to pull, push, and lift, on all terrains. He compared the concept to a cat, which was light but had tremendous climbing ability.

The final design, modified from the British Ministry of Munitions tractors, was simple, compact, and ingeniously featured a motor bolted directly to the transmission housing, an important key to the assembly process. They called it the "three-unit system," which included the rear unit transmission, a central unit that contained the flywheel and clutch, and a front unit to fasten the axle and steering device. The three components could be brought together on a conveyor and bolted together.

The tractor was powered by a twenty-horsepower, four-cylinder Hercules engine. An early plan to adopt the Model T engine was abandoned when designers "became convinced that the kind of tractor we wanted and the automobile had practically nothing in common."[6]

Electricity was generated from a small magneto bolted to the flywheel, ignited by a small amount of gasoline, then switched to more economical kerosene for operation. The tractor weighed 2,500 pounds and featured heavy rear wheel cleats for traction, like the claws of a cat in Ford's comparison. The tractor was a mere five feet high, shorter than the wheels alone of many steam- and gas-powered prairie tractors. Its gray color was a common choice among tractor manufacturers, set off by deep red, sixteen-spoked steel wheels.

Long-anticipated distribution plans for domestic sales were released in March 1918. Henry Ford & Son would begin to sell small allotments of tractors, at least by automobile standards, through a carefully negotiated direct purchase system through state and county war boards, which worked to equitably allocate scarce goods to states during the war, putting tractors on American farms that spring and summer. Ford dealers, knowing little more about the tractor than the public, were excluded from the arrangement.

The tractor, still prevented from using the Ford name, was called the Fordson.

The first Fordson Model F, at last, came complete off the line at Henry Ford & Son's Dearborn factory on April 23, 1918. The name Fordson was pressed into the fuel tank, and Henry Ford and Son cast into the top of the radiator. It was shipped to Ford's friend, well-known botanist and horticulturist Luther Burbank, in Santa Rosa, California. After the machine was unloaded, Burbank looked it over and remarked, "Just like Ford, all motor and no frame."[7]

Luther Burbank's Fordson tractor, the first distributed in the United States. The Metcalf-Fritz photographic collection, BIOS PIC 01:309c, Marian Koshland Bioscience, Natural Resources & Public Health Library, University of California, Berkeley.

Proudly, Ford sent the second tractor to his hero and mentor, Thomas Edison.

The state of Michigan ordered one thousand tractors for resale to farmers, with more than thirteen thousand consigned by June. More than five thousand had already been delivered. That same month, 750 of the state of Minnesota's allotment of one thousand tractors were sold after a two-day demonstration on a thirty-acre plot owned by the lieutenant governor.[8]

The Binford-Kimball Motor Co., distributors for the state of Utah, hoped desperately to have "at least one tractor in Ogden by the first of July."[9]

An early and enticing $200 sales price was now, with the tractor's actual release, a much higher, yet still affordable, $795, minus fenders and belt-pulley. The tractor, like most, was intended to offer a full replacement of the horse, or in this case, three to four horses, but could perform additional farm work, customers were told. "Other tractors now on the market will do these things and do them well . . . But Ford's new tractor is to the tractor world what Ford's auto is to the automobile world . . .

It is cheap.
It is easy to operate.
It burns gasoline, kerosene, or alcohol.
It is simple to construct.
It can be built in vast quantities."[10]

Purchase of a Fordson was not only prudent, but patriotic, potential buyers were informed, alluding to food conservation campaigns and the conversion of factories to military production. "The urgent need for great production and the vital necessity of employing man-saving machinery in farm work is apparent to everyone," advertised an Idaho distributor. "Remembering these immense tasks ahead, the coming of the Fordson is of importance to every farm owner at this time."[11]

11

THE JOHN DEERE TRACTOR

Despite mounting anticipation for the arrival of Henry Ford's tractor on American farms and the expected impact from competing manufacturers, John Deere continued its development efforts. Their program was not a well-kept secret, nor was it intended to be. In recent years, as experimental tractors were tested at the farms of executives William Butterworth, George Peek, and Willard Velie, and on a recently acquired farm adjacent to the Harvester Works in East Moline, a revolving door of mechanics and engineers visited to contribute their expertise. Secrecy was much less important than building a tractor that met the criteria they established in 1912: to meet the qualification of the general-purpose type and "divorcing" the tractor from the plow. Designers consulted locally with the Midland Auto Company on wheel construction; the Associated Manufacturing Company of Waterloo, Iowa, and the Root & Vandervoort Company on engine design; the Kingston Carburetor Company; and a long list of others, some under contract and some just trying to help.

In company publications, tractors appeared in operation with John Deere plows and cultivators. Dr. W.E. Taylor's book, *Soil Culture and Farm*

Methods, included two Deere tractors not yet available for sale, the Dain and the Tractivator.

Despite mounting evidence reinforcing the performance of Joseph Dain's tractor, its commercial viability seemed to diminish monthly as smaller tractors and shrinking prices continued to trend in an increasingly crowded field. Uncertainty around American entry into the war, which seemed more inevitable than ever, further contributed to Deere's lack of commitment. There seemed to be few paths to profitability.

In July 1917, soon after the sudden and debilitating German air raids throughout London that postponed initial consummation of Ford's tractor deal with the British government, George Mixter resigned as John Deere's superintendent of factories and moved to Washington, DC. He would lead airplane inspections for the Ordnance Department, a support arm of the Army. That fall, Charles Melvin, designer of John Deere's first experimental tractor, died during routine hernia surgery. Unusual but not abnormal in a time when a hernia was an oft-recurring issue.[1]

In October, Joseph Dain once again headed to Huron for further field tests. With Sklovsky's D-2 tractor and Brown's Tractivator program both canceled, his chain-driven, all-wheel drive machine was Deere's only remaining design. Following his tractor in the chill and light rain, Dain caught a cold, which was not severe enough to stop work, but eventually "gave him such trouble that he entered a hospital for treatment" in Minneapolis, 500 miles away. "His condition was thought to have improved . . . and his son came to take him home, when a sudden change in his condition proved fatal, death following in a short time."[2]

The deaths of two of Deere's leading tractor designers, coupled with Mixter's departure and Theo Brown's full-time attention diverted to the design and production of ammunition carts and ambulance wagons for the French and American armies, seemed to doom the introduction of the John Deere tractor.

Weeks after Dain's mid-November funeral, Deere's board heard yet another update on Dain tractors in the field. But instead of despondence, there was optimism and an urgency to act. Leon Clausen, just recently transferred to Moline from a Deere factory in Wisconsin, would now lead the project.

❦ ❦ ❦

Forty-year-old Leon Clausen was born in Wisconsin, the third of five children, the first four of whom were boys. Leon's oldest living brother, Frederick, was president of the Van Brunt Manufacturing Company, a maker of grain drills in Horicon, Wisconsin, acquired by Deere in 1911. At the time it was the largest acquisition in company history.

Leon "[led] the usual life of a boy in a small town adjacent to lakes and rivers," earning money mowing lawns and doing odd jobs on surrounding farms. He studied electrical engineering at the University of Wisconsin, then went to work as an electrician's assistant, maintaining batteries and developing lighting solutions for a number of companies. In 1903, he began work as a signal engineer for Bell Telephone.

Meeting with his brother Frederick at a field demonstration for Van Brunt drills, Clausen by chance met George Mixter and was offered a job on the spot "without any definite duties, and I accepted the proposition . . ."

Clausen did a variety of work at Deere before heading to Ottumwa, Iowa, to manage the Dain haying equipment factory. There he stayed until moving to Moline in April 1917 to take over Theo Brown's Tractivator program. On Mixter's departure, Clauson was given responsibility for all manufacturing at Deere, which now included "the manufacture of one hundred tractors of the Dain type."[3]

The first fifty Dain tractors were to be completed by June 1, 1918, and the engines were already on order. Joseph Dain, Jr., "who has had wide experience in the development of the tractor in connection with the work of his father," was to be used "to the fullest extent and in as important a capacity as his experience and abilities permit." A factory report at the end of 1917 confirmed a "desire to build one hundred for sale and further observation. This work will be carried on during the winter and spring, with the object of getting them into the field during the summer of 1918."[4]

Willard Velie, whose automobile line continued to expand, and his older brother Charles from the Kansas City sales branch, pushed for more. In a frustrated letter, Velie reminded Butterworth of the unanimous resolution passed in 1912 to build a tractor plow, observing that "five years and ten months have elapsed" and "our position as either tractor or plow manufacturers is not as strong today as when we started." Deere could not profit from one hundred tractors, he argued, and instead would alienate existing distribution partners. "I cannot refrain from remarking that we should build tractors largely and whole-heartedly, or dismiss the tractor matter as inconsequential and immaterial. Our present course is prejudicial and impotent," he wrote. "I desire to go on record as believing firmly the future of Deere & Company, imperatively and insistently requires immediate action," he wrote, urging greater production of the Dain tractor.[5]

Deere began developmental work on tractor implements well before the break of the twentieth century. At the time they were heavy, massive implements built for steam-powered tractors and early gasoline tractors like the Big Four 30. But the inevitable sale of a Ford tractor in the United States posed significant risk to Deere's leading position in the plow business.

"We are following very closely the development of the Ford tractor and the Oliver plow which goes with it," Dineen updated George Mixter, who retained his position on Deere's board of directors during his wartime service.[6]

Ford gave audience to the dozens of manufacturers designing implements for the Fordson tractor, striking an arrangement with his "good friend" J.D. Oliver for the Oliver Chilled Plow Company's two-bottom No. 7 plow, which would be ready in time for the spring trade. The company's South Bend, Indiana, factory started an aggressive expansion to meet expected demand. The deal with Oliver created greater urgency. Long-time Deere sales manager George Peek, who resigned his position as sales manager to accept an appointment as the Industrial Representative of the War Industries Board, a recently created government body created to standardize industrial output during the war, advocated for a Deere and Ford collaboration. The same board had also arranged for International Harvester's Bert Benjamin

Fordson tractor and No. 7 Oliver plow. Library of Congress, 2016823782.

to visit Dearborn and collaborate on the design of a grain binder compatible with the Ford tractor.[7]

Peek first met with Henry Ford soon after arriving in the nation's capital and shortly after, by telegram, took the "liberty of giving a letter of introduction to you to Mr. H.B. Dineen, Manager of our Plow Works in Moline," vouching for him as of "a very high class and competent plow man."[8]

Ford had no intention to build implements. They were not suited to the assembly line and generated low profit margins. But everyone understood that Ford could change his mind at any moment. "We feel that our plows will be so appealing that they may have some influence on the decision of the Ford people as to whether or not they are going to sell plows with their tractors," Dineen told Peek in mid-February 1918.

Theo Brown shipped two John Deere light, two-bottom plows redesigned for the Fordson, and made plans to visit Dearborn in mid-March to meet with Henry and Edsel Ford, Charles Sorensen, and others. A deal with Ford would certainly cement Deere's future preeminence in the plow business.[9]

⁂

Unlike the Model T, which was now famously available in any color as long as it was black, gray was the color of choice for tractor manufacturers, with some exception. Among the rest, the green, red, and yellow Waterloo Boy tractor stood out.

The Waterloo Gasoline Engine Company was the direct descendant of John Froelich, a "quiet and retiring" man, who built an operational gasoline traction engine in 1892—the word *tractor* had not yet found popular use at that time. Froelich used a sixteen-horsepower stationary engine, mounted on a truck, to create a gasoline tractor that moved both forward and backward; he was later credited as the first to accomplish the task. The following year, Froelich and a sixteen-man crew used his traction engine and a J.I. Case threshing machine to thresh more than 62,000 bushels of grain, primarily wheat. Over the next year, Froelich proved that his outfit could run for half the cost of a steam engine. By the time the Waterloo Gasoline Traction Engine Company was formed in 1893, that cost fell by two-thirds, to $2 a day.

"He is confident that his engine will revolutionize the motive power of threshing machinery and in time will be the successor of steam," reported the *Waterloo Courier* in anticipation of a successful live demonstration on the dirt streets of Waterloo. Four were completed and sold. Two were returned. Eleven more were underway, they shared. "If they only had them, they could sell twenty-five in less than one week." Patterns, coming from Chicago, were late. A farmer in South Dakota ordered one. Curious farmers came from one hundred miles away to see the demonstration on his farm, only to learn it had been cancelled.[10]

Froelich might equally be credited as the first to lose nearly everything because of the farm tractor. Less than two years after its debut, Froelich's partners committed resources to the production of small stationary engines, reincorporating the company without him.[11]

The reformed company introduced the "Waterloo Boy" trade name in 1905. The engine was available under dozens of brand names under license. Making its mark with its redesigned stationary engine, tractor development again began in earnest in 1911, starting with a demonstration of a tractor built in Moline by E.B. Parkhurst, a journeyman inventor residing in Moline at the time. At one point, the Waterloo team met with the Reverend Daniel Hartsough, who was then looking for a builder for a new tractor he called the Bull. They declined in favor of their own design.

About twenty model LA tractors, a four-wheeled tractor with a two-cylinder, opposed motor, were first sold in 1913. A redesigned, single-speed tractor, the Waterloo Boy model R, capable of pulling two plows, was available by the spring of 1914. More than one hundred were sold over the next year.[12]

The Waterloo Boy tractor was a regular participant in the National Tractor Demonstration and had carved out admirable single-digit market share in recent years. During final Dain tractor tests in the fall of 1917, Deere's representatives monitored a Waterloo Boy, noting that ". . . it can burn kerosene very successfully. This point is important for the reason that if the time should come when gasoline is to jump to 30 or 40 cents per gallon, and we did not have a successful kerosene burning motor, we would be up against it."[13]

War restrictions had already driven the price of steel and other raw materials upward of 50 percent in recent years, and input costs for farmers were of growing concern. The average price of gasoline would in fact rise to nearly 30 cents the following year, only two-thirds the cost of a pound of coffee, but 7 cents more on average than a gallon of kerosene.[14]

Deere's connection to the Waterloo Gasoline Engine came about through the efforts of Otto Borchert, the son of a Milwaukee brewer, Frederick Borchert, not to be confused with the Fred Borchert who would eventually be acquitted of murdering the then husband of his future wife. Otto's great-uncle Ernst was president of the Pabst Brewing Company.

Borchert was a born salesman, adept at connecting people to opportunities, first for the Wisconsin Telephone Company, then for Julius Andrae & Sons, one of the "largest electrical jobbing institutions in the United States." In 1917, he brokered the sale of the Union Electric Company of Chicago to I.C. Elston of Waterloo "at a price said to be several million dollars." He matched several other electrical companies with buyers as well. While in Waterloo he befriended George B. Miller, president of the Waterloo Gasoline Engine

Company. When Miller began to talk of retirement, Borchert "immediately tendered his services."[15]

At a board meeting in late January, Frank Silloway, whom Borchert considered a "very smart man," brought Deere information about available options on the Waterloo Gasoline Engine Company. He added to the growing sentiment that Deere's salesmen and dealers thought implement sales were at considerable risk without a tractor and that their plan to build up to one hundred all-wheel-drive tractors had not fully addressed the issue on the necessary scale. William Morgan, the former International Harvester sales manager now running the John Deere Harvester Works, agreed but didn't think they had enough information from the field to move forward with anything. He suggested letting the sales force decide "whether they want a tractor, what kind of a tractor, and then let the factory put it in shape to manufacture." In the meantime, they shouldn't waste any more time figuring out what type of tractor to build. "It is going to be a sort of evolution until we reach a standard," he said. "I do not think there are any of the tractors that are entirely satisfactory now as they should be or will be."[16]

Besides stirring interest from Deere, Borchert shopped the Waterloo Gasoline Engine Company to the Allis-Chalmers Manufacturing Company, a small but emerging company with a few years of experience in the tractor market. Sears & Roebuck, which had grown from its first jewelry and watch catalog in 1888 to a mail order catalog powerhouse with 1918 net sales of nearly $200 million, showed interest as well. A tractor would complement existing offerings for tractor engines, cushions for tractor seats, oil, even a large variety of plows, planters, potato diggers, manure spreaders, and dozens of farm implements from maker David Bradley.[17]

Frank Silloway headed to Waterloo, followed by engineer Max Sklovsky a few days later, to investigate the opportunity in person. There, they met Louis Witry, the technological genius behind the company's product line.

Waterloo native Louis Witry, born in 1879, started a five-year apprenticeship at the Illinois Central Railroad at the age of fifteen. He joined the Waterloo Gasoline Engine Company in 1897, just a few years after the company parted ways with founder John Froelich. Automobile production, a common rite of passage for engine and tractor-makers, made a brief appearance as well.

Witry often took to the racetrack to test his designs. In 1902, three years after the debut of the Home Park half-mile track in Waterloo built for horses competing in trotting, pacing, and racing, he won the track's first automobile race. His car broke down midway through the first lap, but after a few minor repairs, he went on to win the five-mile race by a full lap. The races at Home Park had a small following to start; then attendance exploded. In 1906, they set a record with more than eleven thousand spectators. The Labor Day race the following year brought a crowd of twenty thousand. Unfortunately, high attendance created a high demand in local real estate, and the track's owners found greater profits splitting the eighty-acre property into lots and selling them off one at a time, ending the Home Park races.[18]

In 1918, Waterloo's 128 factories employed six thousand workers, forming "the foundation stones of the city's prosperity and rapid growth." Four companies were building tractors and engines: the William Galloway company, the Iowa Dairy Separator Company, the Inter-State Tractor Company, and the Waterloo Gasoline Engine Company. More than forty businesses were supporting the automobile industry through parts or service. Galloway, who had sold everything from stationary engines to manure spreaders to washing machines, was just one of many local entrepreneurs who had separately tried their hands at automobile manufacturing as well, some prior to Ford's introduction of the Model T.

Impressed by Witry, Silloway found a "good factory" with room for expansion and "up-to-date machinery," including three modest buildings made of cement block and timber on forty acres, with "ample room for expansion." More than $250,000 had been put into new buildings and equipment in the previous year, including a foundry, which was not yet complete. Even more intriguing, Silloway was told of a new tractor, far along in its design. The new Waterloo

Boy model N would be a two-speed, two-cylinder tractor, but its design would be "more compact, improved in appearance."[19]

Sales of stationary engines were in a three-year slump, but projections showed great potential for steady future growth. Tractor sales reached 2,762 in 1916, surging to 4,558 in 1917, and with even better prospects for the following year. The Waterloo Boy was a ready-made alternative to Joseph Dain's more expensive all-wheel-drive tractor.

The Waterloo Boy, the brand, not the tractor, came to life as an actual Waterloo boy—a cherub-faced farm boy with red rosy checks, a straw hat, bib overalls, and a shaft of hay dangling from his mouth. The tractor had appeared in a number of John Deere field reports over the previous years, with mixed results. But it featured the durability, ease of service, and ruggedness Deere wanted in a tractor. It burned kerosene, Deere's fuel preference due to costly and unstable gasoline prices during the war—and expected volatility after. Deere forecasted post-war price pressures on all fronts to manufacturers. For a company focused on steady, incremental growth, volatility was the enemy. Butterworth, Deere's fiscally conservative leader, told the *Boston News Bureau* later in 1918 that all near-term purchasing would be done "in hand-in-mouth fashion" and that dealers and farmers would wait for cheaper prices due to overproduction. He predicted lean times for farmers and, as a result, the entire agricultural equipment industry.[20]

The Waterloo Boy had a two-cylinder engine, which could be built more cheaply than its own four-cylinder, all-wheel-drive design and was durable and used fewer parts. It was economical, serviceable, and purpose-built to "pull hard all the time." Silloway left his visit of Waterloo, Iowa, with a favorable view: "I believe that, quality and price considered, it is the best commercial tractor on the market today." The Waterloo Boy would give Deere "a satisfactory tractor at a popular price, and not a high-priced tractor built for the few."

Silloway disagreed with William Butterworth's assessment of the business environment and still expected the implement business to remain steady in 1918 and 1919, and "by 1920 the war should be drawing to a close, the period of transition would be upon us, and I believe that none of us know what it will bring." To that end, he thought Deere could wait to get into the tractor business. "All it would require is money for a plant and machinery," he said. But "this Waterloo proposition presents the subject in a new light," and as a result, Deere had "an opportunity to, over night, step into practically first place in the tractor business."[21]

Step in they did, swiftly approving the $2.25 million acquisition of the Waterloo Gasoline Engine Company. Importantly, the three-plow Waterloo Boy would also not compete with the two-plow Fordson.

Silloway thought the Waterloo Boy could surpass International Harvester and secure a strong second-place showing behind the Fordson. Charles Deere Velie agreed. "I am more than satisfied we have made the best move Deere & Company has ever made," he wrote Butterworth. "I believe if we handle this proposition right, the Waterloo Boy will be to the tractor trade what the Ford car is to the automobile trade. Of course the Ford tractor will take first place, but if we can take second place that will be good enough for us."[22]

Waterloo was a good fit because of "its well-known adherence to the highest quality of its output," an internal company magazine touted, and that the "products of the Waterloo Gasoline Engine Company will add new lustre to it."[23]

Model N Waterloo Boy tractor pulling a two-bottom plow. Courtesy John Deere.

The model R, in production since 1914, was still being phased out and replaced by the new model N, which began production in December 1916. Both were sold in 1917 and into 1918, and would continue until the R parts were used up, with a total of eight thousand model Rs built over a four-year span, including exports. A record number of tractors, about four thousand, were built in 1917, a more than 40 percent increase over the previous year.[24]

The Waterloo Boy N, the "size demonstrated by experience to be the most successful on the average farm," pulled three 14-inch plows in "ordinary soil" and two binders in heavy grain. It was constructed with a six-inch channel steel frame, a patented kerosene manifold for greater fuel efficiency, and a fully protected crankcase to protect bearings from dirt. Two forward and one reverse "automobile style shifts" provided a top speed of three miles per hour. The tractor was measured at a height of five feet, three inches, and a length of eleven feet.

"I did not use more than two gallons of kerosene to the acre plowing at good depth with three-bottom, 14-inch John Deere Plow," submitted Charles Carlson of Stromsburg, Nebraska, in a testimonial.[25]

Deere's acquisition of the Waterloo Gasoline Engine Company in March 1918, a month before the first Fordson emerged from the Ford factory, was the largest financial transaction in the history of the city of Waterloo and included more than forty acres of property. In the weeks after, Martin Fleming, a thirty-six-year-old Swedish machinist at the John Deere "tractor factory," referring to the Marseilles Works in East Moline, traveled to Garrett Park, Maryland, with William Butterworth and other Deere representatives. In tow was the first of the all-wheel-drive tractors that Joseph Dain had agonized over and fought for, but had never seen in final form. Fleming's audience included the secretary of agriculture, testing tractors for government purchase.

In a matter of weeks, John Deere now had two tractors to offer. Materials were on hand to build "up to 100" of what was simply called the John Deere Tractor, but with the Waterloo acquisition complete, the future of Joseph Dain's tractor was again unclear.[26]

12

"FORD LIKES A SUCCESS"

Agricultural equipment manufacturers felt like they were at last catching up to the automobile industry. The Society of Tractor Engineers consolidated with the Society of Automotive Engineers and the American Society of Aeronautic Engineers, pooling resources and expertise to the advantage of all. "This the tractor men needed badly," rationalized International Harvester president Cyrus McCormick, Jr., admitting that they had much to learn about material strength, the heat treating of steel, and refinement of design areas, all areas well understood by automotive engineers. Henry Ford's early adoption of vanadium steel, lighter, stronger, and more precisely machined, proved the point. The prospect of this newfound knowledge offered hopes of advancement and innovation in the tractor field.[1]

Members of the National Implement and Vehicle Association closely monitored Ford's tractor distribution arrangements with each individual state. At their meeting in Chicago, they drafted a letter to dealer associations nationwide to learn how they were guarding against Ford's distribution plans. They also began to investigate legal action and to reach out to state governors. Collectively, they considered the Ford arrangement, like the initial arrangement with the British government a year before, to be suspicious and potentially counterproductive for the industry, for "if the tractor does not prove to be satisfactory" it would "create a severe blow to the tractor within the states."[2]

Theo Brown and two other Deere representatives got their long-anticipated audience with Charles Sorensen, who had been in on the Ford tractor since the beginning, and Henry Ford himself, in mid-April, the first of at least six separate visits Brown made in the spring and summer of 1918. The Ford farms now included more than seven thousand acres, more than half of that under cultivation, running between twelve and seventeen tractors at any time.

A tour of the tractor factory left a deep impression. Ford told them "he spent a lot of money fast on his tractor plant," $5 million, he confided. They were already finishing 125 tractors a day, and soon that number would be a thousand. "It was easy to manufacture if you only build one thing," Ford shared.

Both Sorensen and Ford liked the Deere plow "much better than Oliver . . . ," Ford offered. In fact, the next time he saw his "good friend" J.D. Oliver he was going to tell him that he "has to go some to equal the Deere plow." Ford was always negotiating.[3]

A week later, armed with information from Brown's visit, the manager of each of Deere's factories met in Moline. Among the many discussion topics was the adaptation and design of plows and other implements for tractor use. One issue, with a singular response, had become clear:

1. The Ford tractor may herald the elimination to a considerable extent of horse-drawn implements.
2. Every factory organization with a hearty support of the General Company should develop their individual tools as to make them adaptable to this tractor.

As a reminder, Frank Silloway "called attention" to the fact they had just purchased the Waterloo Gasoline Engine Company and "suggested that we adapt the various tractor tools in our line particularly to the Waterloo Boy tractor in addition to the more particular tools for the Fordson tractor."[4]

One thing most tractor manufacturers of the day could agree on was that head-on competition with Ford was competitive suicide. Most resolved to supplement the Fordson, whether their implement was approved by Henry Ford & Son or not, and take advantage of the volume Ford promised. Baker Manufacturing in Springfield, Illinois, offered a reversible, one-man plow that could clear a six-foot-wide path. An external steel frame could be mounted directly on the Fordson. A steering column ran perpendicular to the tractor body, above it, with a dedicated steering wheel to move the plow. A v-plow was available to move snow to both sides at the same time.

The Wehr Company, of Milwaukee, introduced a 1,265-pound road grader attachment that bolted to the tractor chassis. With the front wheels of the Fordson removed, the entire outfit weighed in at more than four thousand pounds.

Even International Harvester continued work on a binder and other implements for the Fordson.

Regardless of Deere's acquisition of the Waterloo Gasoline Engine Company, coupled with production of Joseph Dain's tractor, a John Deere plow for the Fordson also seemed imminent even in 1918. But the companies were also still irreconcilably apart on the terms of distribution. When the sale of tractors through county and state war boards expired, Ford intended to push tractor and implement sales onto his own expansive dealer network. Deere's team thought the strategy was a mistake. Besides convincing Ford of the superiority of the Deere plow, Dineen and Brown now had the much more difficult task of trying to "convince them that we are interested and will take care of plows for their tractors through our regular trade channels."[5]

Before their return to Dearborn to discuss the test results that spring, Dineen learned that the state of Michigan purchased one thousand Fordson tractors and one thousand Oliver plows at a price of $750 and $100 each, all cash, respectively. The Oliver No. 7 two-bottom plow was the first Oliver offering in the Fordson catalog, but Dineen was undeterred. The plow was too heavy, he reasoned, convinced that decades of tested and proven plow design, built heavy for rough work, was outdated. Plows for the Fordson needed to be lighter, like the Fordson itself.

In the United Kingdom, Oliver plows, which were fully endorsed and recommended, were receiving complaints in some circles. The Oliver company thought the concerns were due to "prejudice" or "misuse," which was the "result of ignorance."[6]

Work on a plow for the Ford tractor continued at Deere. To achieve their weight reduction goal, Ford arranged for a supply of vanadium steel from the Central Steel Company in Massillon, Ohio, the same lightweight, durable steel used in the Model T. Ford often said that he learned about vanadium steel in 1905 after an automobile race pileup in Palm Beach. After the race, he walked around the site and "picked up a little valve strip stem. It was very light and strong," he recalled. Unable to find anyone who could tell him what it was made from, he gave it to his assistant. "Find out about this," he told him. "That is the kind of material we ought to have in our cars."[7]

Perhaps there were bits of truth to the story, but he was truly, and factually, convinced a year later after a visit by an English metallurgist named J. Kent Smith, who demonstrated the material to Ford and Sorensen in Detroit. Ford told *Harper's Weekly* that vanadium "imparts qualities to steel which are little less than magical."[8]

In April, Theo Brown, recently promoted to director of the experimental department at Deere, and others brought their experimental plow to Fair Lane, the fifty-six-room Ford estate near the River Rouge plant, for Ford to personally test.[9]

The two-bottom Deere plow weighed 416 pounds and was tested directly against the 590-pound Oliver. "A girl was able to take hold of the rear brace and lift the plow around," favoring the Deere. It was stiffer and built with a better axle, allowing it to be transported at higher speeds moving from field to field. The Fordson could move across dirt roads at ten to twelve miles per hour, and the plow had to withstand the ruts and bumps at high speed without the plow dropping or sustaining damage. Deere's plow also had half as many parts as the Oliver, used rivets instead of bolts, and matched the Fordson's "trimness and cleanness of design."

At one point, Ford himself operated a Fordson tractor coupled with an Oliver plow. When it failed, he called for a new Oliver, only to get the same result.

Deere's Theo Brown was key to ongoing negotiations with Henry Ford. This portrait was taken in 1918. Courtesy John Deere.

Ford was "beginning to realize that the tractor cannot be used with obsolete tools," and that to make the Fordson a "commercial success it will be necessary for the implement manufacturers to produce satisfactory implements of all kinds and a large number of them," Brown reported.

In their frequent visits, not two years removed from Brown's initial starstruck meeting with Ford at the National Tractor Demonstration in Fremont, Nebraska, he had developed critical insights into Ford's operations. Implements had to be "fool proof" like the Fordson. To make it so, they sent "greenhorns with the new tractors to try out on his farms." They tested all types of implements from dozens of manufacturers. If it did not work, Ford himself would not see it.

"Ford likes a success," Brown reported, "but won't look at a failure."

Soon after the plow tests, Ford's brother-in-law, Edgar R. Bryant, who had the rights to sell Fordson tractors in the state of Michigan, ordered one hundred plows directly from Deere. But as hard as they had worked to get to this point, they told Bryant he would have to wait. The plow was not ready, they contended.

Ever optimistic, Frank Silloway prepared to sell fifteen to twenty thousand Fordson plows a year and spearheaded efforts to outfit the Plow Works for production, though none would be released until they came to an agreement on distribution. "In the meantime," Floyd Todd shared with Bryant, "we have not determined what our sales policy will be and are not now in position to accept your order. The only thing we can say to you is that after the plow is developed, if we determine to market it through other than our own channels, we will give you consideration for the state of Michigan."[10]

Lacking the scale of a Ford, or even an International Harvester, the protection of distribution channels was of paramount importance to Deere. And Ford's position, outside of the current arrangements with state governments in support of war production, had yet to be clarified. Larger firms had an established network of wholesale distribution centers called a branch house. These sales branches operated sub-branches and transfer houses, a regional distribution network that stocked unassembled implements and parts and provided marketing and logistical support to their territory.

John Deere operated twenty-nine branches and sub-branches in the United States and Canada, from Atlanta and Baltimore, to Dallas, New Orleans, San Francisco, and Portland. Comparatively, International Harvester ran its operations through ninety-three branches in the United States and seventeen in Canada.

Dealers literally and figuratively had their feet on the ground, and through their daily interactions they provided a crucial customer connection that could never have been achieved from the corporate office. The sales networks of farm equipment manufacturers had evolved in the era of consolidations and mergers and were often the strongest indicator of a company's health.

Advanced price lists also began to be published in this period, with sales branches placing orders based on forecasted demand from their territories, also supplementing their own lines with equipment from local manufacturers when necessary. Ultimately, dealers, hardware stores, or retailers would purchase their allotment at wholesale cost, then sell directly to customers at a suggested retail price and provide support after the sale.

Every territory and farm within that territory was different, and over time, dealers developed deep relationships with their customers, understanding their particular needs and goals. Companies like International Harvester, John Deere, the Moline Plow Company, J.I. Case, Avery, and others with a deep catalog had also moved toward exclusive licensing with their dealers, carrying only their full lines.

As testing continued on what Deere now designated the No. 40 Plow, the topic of distribution was raised with Ford on several occasions, but he tended to deflect the question. "Mr. Ford said he did not think we needed to worry about sales if the plow was right," Brown wrote in his journal that summer. They discussed the topic with Joe Galamb, "who had a diarrhea of ideas," according to Ernst Kanzler, and Charles Sorensen. In the course of a two-day visit in July, they learned that the Ford policy was actually no policy at all. "The distributor simply buys tractors from Ford; he has no given territory nor is any retail price given for him to resell at. It is not necessary for the distributor to handle implements, but he probably will want to and Ford will not stop him."[11]

Ford's actual treatment of automobile dealers, they thought, told a different story.

Upon their return to Moline, a new set of plows was shipped to Detroit for further testing.

Brown, C.C. Webber, the manager of Deere's Minneapolis sales branch, and Charles Deere Velie from the Kansas City branch, met in Valley City, North Dakota, to test a Fordson and an updated plow built with vanadium steel. It tested well, but their new Waterloo Boy tractor raised concern. "The Fordson did so nearly what the Waterloo Boy did that it makes one think that something should be done to the Waterloo Boy to increase its power," Brown confided in his journal.[12]

Velie was more optimistic, skeptical the Fordson would hold up under tough operating conditions. The tractor was too light and "the traction will prove entirely unsatisfactory," he predicted. "I hope he [the Fordson] is not in that class," he further rationalized. "If he can do as much work per day pulling two plows and running faster than the rest of us, as we can pulling three bottoms at lower speed, he will indeed be a dangerous competitor."[13]

Despite their anxiousness to sell the Waterloo Boy, Deere decided to delay its introduction in honor of pre-existing contracts with jobbers and independent dealers. Those relationships, now or later, would be important. Deere issued a bulletin to its branch houses to clarify, giving direction that if dealers asked about the sale of Waterloo Boy tractors or engines, "the only thing for you to tell them is that you have not taken over the sale of the tractors or gasoline engines as yet." Existing contracts between the Waterloo Gasoline Engine Company, jobbers, and dealers expired December 31, 1918.[14]

"It is our intention . . . to sell the Waterloo Boy Tractor through our established John Deere dealers and thereby strengthen the prestige of the John Deere line with our dealers and thus increase our trade on the John Deere line, and the power machinery in particular," they informed the branches in July. "We bought the Waterloo Gasoline Engine Co. for our dealers and consequently we should, as a general rule, handle the Waterloo Gasoline Engine business so as to give the local agency to our dealers and to give them the same territory on the Waterloo Boy Tractor as they have on the balance of our line . . ."[15]

Deere allocated $50,000 for the Waterloo Gasoline Engine Company advertising program for the twelve months beginning April 1, 1918. The Waterloo Boy tractor was to be positioned as the "best and most efficient tractor" on the market for farmers inclined to buy a tractor. Advertising would be executed in "such a way to call the attention of the farmers and dealers to the 'Waterloo Boy' tractor in comparison with other tractors and eliminate from our advertising all matter of an educational nature that tended to boost the tractor at the expense of the horse."[16]

The acquisition would "increase the prestige of the John Deere line with our dealers, and we must so handle this tractor proposition to make it count for the most in the sale of our line and the establishment of John Deere Full Line representation in the town."

Dealers were urged to "arrange their affairs as to get into the tractor business." If they "neglect this important branch of their responsibility" they might lose out to the automobile dealer in their town, who would in turn become "the tractor man in their town . . ." A dealer "must realize that he cannot hold this position to which he is rightfully entitled unless he recognizes and fulfills his responsibilities . . . to be an active and energetic tractor and tractor implement salesman, in addition to the regular responsibilities he has had in the past of supplying the farmer's needs on all the implements he requires."[17]

The Fordson factory built 791 tractors in the six days ending July 13. They made 135 on the fifteenth, 117 on the sixteenth, and 115 on the seventeenth. Production continued to grow, according to numbers secured by Harold Dineen.

"We succeeded in getting before Mr. Sorensen the fact that we had a problem to decide as to the channel through which we should market the No. 40 plow," Dineen reported. "He agreed that this problem was a real one and that it was something we had to decide ourselves."[18]

During their visit, arrangements were made for Ford, Sorensen, and others from Ford to stay together for two, perhaps three days, during the National Farming Demonstration in Salina, Kansas, in just a few weeks' time, in August 1918. In the meantime, Galamb and Farkas traveled with Theo Brown from Detroit to Moline to learn firsthand about the production of John Deere plows and inadvertently give the local press reason to speculate about the growing relationship between the two companies.

13

TRACTOR CITY

Despite educating millions of customers on the latest agricultural equipment technology over the last four years, a national demonstration circuit and its continuation were hotly debated throughout 1917. And while Henry Ford shipped tractors to Great Britain, American farmers were buying a wide variety of tractor styles from more than a hundred different manufacturers. "The elimination of the freak tractor has been a very common remark," wrote chairman J.B. Bartholomew. The events showed all in attendance that "there are successful tractors on the market." Admittedly, they could do more to show spectators which tractors "could handle the work," but substantial progress could not be denied. Endurance tests were proposed for the following year.[1]

Newcomers in need of the exposure, like the Gray Tractor Manufacturing Company, makers of a unique "drum drive" tractor, continued to praise the events, while mainstays thought the expense was beginning to outweigh the benefit. International Harvester thought the events had run their course. After another showing in Fremont during the summer of 1917, it was clear that they "lacked the old-time interest," mused the editors of *The Harvester World*, a magazine published for International Harvester employees and dealers. "It seemed a little antique, like a style of clothes that has gone out-of-date, or the habit of sending picture postal cards." Perhaps the national demonstrations should now go to China or Russia, a company representative mused. In the United States, the benefits no longer justified the trouble and expense. The show planned for Salina, Kansas, in August 1918 might be the last one, they conjectured.[2]

By the time of the show, that seed had been firmly planted. The National Tractor Demonstration continued into 1918, but it was well circulated that the event in Salina "might be the last ever held in the United States," the *Abilene Weekly Reflector* offered in its headline, just days before the event was to kick off.[3]

After seven years in Fremont, one of the few repeated stops, Salina, offered a much-needed change of scenery for manufacturers and customers alike. Over the years, the Kansas Pacific Railroad transformed Salina, located in northwest Kansas, from an outpost to a town, ushering in a period of settlement that brought churches, schools, a newspaper, saloons, and, eventually, notoriety for its growing cattle trade through the early 1870s. When that moved west, Salina became wheat country. In 1903, most of the town was consumed by the flood waters of the Smoky Hill River, then rebuilt.

The transformation of the tractor and the industry around it was evident everywhere. The average weight of tractors on exhibit, calculated per brake horsepower, was a mere 297 pounds, compared to 681 pounds in 1908. "No more trumpeting over the prairies with grimy perspiring operators clutching a mad wheel that turned as hard as a tugboat in a gale," offered the *Implement & Tractor Trade Journal* in their review of the event. "Today, we have the tractor with parlor manners."

New event chairman A.E. Hildebrand said the demonstration would cost at least $1 million to hold, "but it is worth a half billion to the nation in education towards increased food production, conservation of manpower, and the utilization of the country's energies."[4]

Another testament to the transformative nature of power farming was the growing number of female operators running tractors at the event, a common tactic meant to demonstrate ease of operation. The stunts also affirmed the capability of female farmers. Mrs. W.H. Line ran an Avery motor cultivator for much of the week, "clad in a smart 'tractor habit' of khaki," knee-high boots, and a waist-length jacket, "which they do say came from Marshall Fields." International Harvester recruited two girls from a local college, Miss Cecelia Dorney and Miss Bernice Clark, to operate some of its tractors during the week.

"Tractor City," as they commonly called the well-organized patchwork of tents, exhibits, and sandwich and soda pop stands at the demonstrations, impressed one visitor enough to compare it to the midway at a world's fair. Exhibitors this time were organized by a random drawing, a "more fair method," instead of alphabetical as in years past. Deere pulled the number seven spot, International Harvester number thirteen. Fordson was near the end, number thirty-four of forty-four spots.[5]

For the first time, a broader line of farm machines was permitted to give a more complete picture of the impact of power farming both from the tractor drawbar, which had always been demonstrated, and now the belt, which included demonstrations of threshers, saws, and other equipment. International Harvester, which sold everything, had lobbied for the change.

One hundred and twenty-four companies sold nearly 63,000 farm tractors in the United States in 1917. There were mainstays still. International Harvester, which continued to manufacture truck bodies, ammunition carts, artillery shells, and more for the Allied armies, still built a record 16,101 tractors, a market-leading 26 percent of industry sales.

Harvester brought its typically impressive full line, which included somewhere between ten and fifteen tractors—newspaper accounts weren't overly concerned with exactness—several motor cultivators, and a fleet of plows, harrows, spreaders, and other equipment, including a threshing machine. In the field behind their tent, Harvester hearkened back to their modest tractor origins, running a 1909 model tractor continuously to give "some idea of the life of a tractor."

Prior to the event, Harvester offered a private demonstration of its Mogul 10-20, "an everyday power producer on the farm," at the C.H. Leep farm in North Topeka.[6]

Their exhibit did not tip customers off to the internal changes underway, stemming from their ongoing legal battle with the Federal Trade

Commission. Soon after the demonstration, Harvester dropped its pending appeal in the United States Supreme Court. In the time since the original anti-trust charges were lodged in 1912, Harvester had twice appealed and was waiting to begin a third until the United States entered the war in 1917. Now, "under the pressure of war conditions" according to president Cyrus McCormick, Jr., Harvester had agreed to dismiss its suit. In a statement, he opined that the initial suit was "based not upon any wrongful practices or injurious acts, but upon the company's alleged, but unexercised, power to dominate the agricultural implement trade."

With the case finally settled, Harvester began the process of selling off three of its five harvesting brands—Osborne, Champion, and Milwaukee—and two of its harvesting factories, which it was ultimately allowed to keep. They were so out-of-date that the new buyers didn't want to be straddled with the costs of updating them. Rockford's Emerson-Brantingham Company, which now benefited from "being able to enter the field without loss of time in experimenting," acquired Osborne, while B.F. Avery & Sons in Peoria, Illinois, acquired the Champion line.

More concerning was the court-ordered dismantling of Harvester's massive dealer network and the expected loss of almost five thousand dealers over the next two years.[7]

J.I. Case, Avery, and Rumely were other well-known names in Salina among a field of new companies formed from recent consolidations and partnerships. Other representatives walked the grounds in search of investors, or simply hoping that one more show would provide the means to another year of survival. There were debuts, including the Port Huron Engine & Thresher Co., the R&P Tractor Co. 122 "pad tread" tracked tractor, and the Hession Tiller & Tractor Corporation's Hession farm and road tractor, which they would soon unabashedly advertise as "the sensation of the Salina demonstration." It surely was for the sheer novelty of its primary feature. Its rear wheels could be removed and replaced with rubber-tired road wheels in only thirty minutes' time.[8]

The Bull Tractor Company, which had seemingly overnight converted horse farmers with its small Bull tractor in 1914, brought four of its updated Big Bull tractors. A distribution deal with the Massey-Harris company sent Bulls to Canada beginning in 1917, but the company was now in a fight for its survival. A year earlier, a merger with the Whitman Agricultural Co. of St. Louis fell through, and now it was near a merger with the Madison Motors Corp. of Madison, Indiana. The deal was finalized in September 1918, with plans to move tractor production to Madison.[9]

The J.I. Case Plow Works Company, not to be confused with the J.I. Case Threshing Machine Company, brought fifteen Cub Jr. 15-25 tractors, one Cub 26-44, and a variety of disc and moldboard plows and harrows. Its tent was literally a glowing beacon; an electric dynamometer, driven by a Case 10-18 tractor, illuminated the tent, not to mention the giant, trademarked globe and eagle, which "could be seen for a long distance." The company had come a long way from the sentiment of vice president F. Lee Norton, who told dealers ten years earlier that "any salesman heard praising the gasoline tractor will be fired on the spot." The Wallis was "a masterpiece of tractor genius," advertisements claimed. Howard Coffin, head of the U.S. Aircraft Board and designer of the "famous Hudson Super-Six motor car," bought a Cub for his personal farm while on site. The J.I. Case Threshing Machine Co. of Racine, Wisconsin, which was in no way "interested in, or in any way connected or affiliated with the J.I. Case Plow Works, or the Wallis Tractor Company, or the J.I. Case Plow Works Co.," exhibited an entirely separate line of tractors in five sizes, the 10-18, 10-20, Type "A", 15-27 and 20-40.[10]

Case added Coffin to the list of "celebrity" customers, which included chewing gum baron William Wrigley; King Gillette, "maker of safety razors"; and H.J. Heinz, "famous because of pickles and the '57 Varieties.'" The king of Spain made the list as well. Less notably, George Selby of Abilene won a Stetson hat, his prize for being the first Case distributor to sell a tractor at the event.[11]

❦ ❦ ❦

During his last visit to Dearborn, Theo Brown had been told by Ford that he "would do anything to the tractor to make it adaptable—add lugs or anything else," to bring Deere's work on a plow to a conclusion. At the very least, Brown expected another successful round of negotiations during their week together in Salina.

Deere's Harold Dineen and Theo Brown had plenty to think about as they waited patiently at the station for the arrival of the contingent representing Henry Ford & Son. When the steam locomotive finally came to a stop, Henry Ford found his way to the rear and stepped off by himself. "I suppose the reason being he wished to avoid as much publicity as possible," wrote Brown, who quickly met him but was beaten by a reporter. Ford told her, "I bet a dollar I know what you want," then stepped down and over to the Deere car that was waiting. Presumably it was a Model T.

The parties spent most of the week together, well beyond the two or three days the Deere team hoped. At one point, the companies competed in a spirited baseball game "with real pitching." The Deere squad lost seven to six.[12]

Henry Ford & Son, with thousands of machines already plowing fields across the United States after only five months of production, brought fifteen Fordsons, "all of one size," to the event. Noticeably, still with no deal in place, they were paired with Oliver two-bottom plows, discs, and pulverizers during demonstrations. Of the forty-one tractor manufacturers at the show, twenty-five of them had partnered with Oliver. Most of those tents included an Oliver representative, clad in white from head to toe.

The Oliver Chilled Plow Company and recently the Grand Detour Plow Company were the only two willing to agree to Ford's terms and sell their implements direct through Ford dealers. Neither had a tractor of its own.[13]

John Deere's Kansas City sales branch and a local distributor, the P.J. Downes Company, brought twelve Waterloo Boy tractors, which with its deep green frame and fenders, red wheels and engine, stood out from the gray and dark red offerings from other manufacturers. Decals of the Waterloo Boy, straw hat and all, graced the backsides of the rear fenders, as if watching the plows pulling from behind.

Deere's dealer erected a gleaming white tent, 80 by 120 feet, framed out with sixteen-foot-wide John Deere signs, one on the front and one on each side, with patriotic bunting draping the tent and American flags flying high into the sky. The tent's roof featured a deer leaping over a log. John Deere's son and Deere's second president, Charles, purchased more than a dozen of the copper sculptures at the World's Columbian Exposition in Chicago in 1893. The statues, seen on the rooftops of John Deere buildings across the country, on parade floats, and at farm shows nationwide, forever tied the leaping deer to the John Deere brand.

Once inside, attendees checked their bags, the staff ridding visitors of the responsibility of their personal belongings while they watched the motion picture films shown each afternoon, or, if they preferred, listened to the patriotic music played on the Victrola. A pen of live deer was of "special interest to the younger visitors." If you needed to talk business, staff would watch your children, and had "ice water on tap," a big hit with temperatures reaching over 100 degrees.

In the trade press, Deere pushed its longevity as a company, a long-standing relationship with farmers, and a deep sales and dealer network. A two-page spread advertisement called out the "sweeping demand for Waterloo Boy tractors now, which must be met." It was backed by "ample capital" by an "established reputation," and an "aggressive sales organization and long established responsible manufacturing concern, which insures [*sic*] permanent business."[14]

But the real star of the show for Deere was the Waterloo Boy model N tractor, which demonstrated its merit pulling tractor plows, disc harrows, and grain drills. Visitors were given straw hats to protect them from the sun as they walked the exhibition grounds or were shuttled in three John Deere farm wagons pulled by Waterloo Boy tractors.

"The award for the most elaborate, largest, and most artistic exhibit tent at the Salina tractor show will undoubtedly go to the John Deere Plow company of Kansas City," wrote the editors of one Kansas City newspaper.

Willard Velie, automobile and truck manufacturer, and perhaps Deere's strongest tractor advocate, was there as well, but on business of his own. In protest over his waning authority, he had recently offered his resignation from Deere's board of directors. "I am under the impression that I am Chairman of the obsolescent Executive Committee of Deere & Company," he wrote Butterworth. "In this case, please consider this my resignation of such, as a member

The Waterloo Boy model N made its debut as a John Deere tractor in Salina, Kansas, August 1918. Courtesy John Deere.

of the Committee as well. I will remain a Director of the Company until I shall have disposed of my holdings." He was not in a hurry to leave. The process took three more years.[15]

Velie brought his own tractor, the four-cylinder, 12-25 Velie Biltwell, which he had debuted the year before at a smaller event in Kansas City. Ten tractors were built in 1917, but plans for a factory on fifteen acres in East Moline, just north of the John Deere Harvester and Marseilles factories, were now postponed. Wartime orders for trucks pushed the Velie automobile factory in Moline to capacity. "Rushed work on the trucks for the government and other orders is keeping the Moline factory exceedingly busy," a local paper reported on a massive order by the French for two thousand Velie trucks, "and it is thought that it is due to this reason that the tractor plans have been abandoned for the present." By year's end he would manage to build only five tractors.[16]

❧ ❧ ❧

In head-to-head competition at the demonstration, the Wallis Cub excelled, followed by the Waterloo Boy, a Parrett 12-25, which "speaks for itself" according to literature. The Fordson finished fourth. Comparatively, the three-plow Waterloo Boy did 50 percent more work in a day than the two-plow Fordson. "That was the point that was worrying me more than anything else," wrote Willard's brother, Charles Deere Velie of Deere's Kansas City branch, who watched the competition. "Now, we can go ahead and sell our tractor for 1919 now that we are warranted in asking our price of $1250—and we can sell all we can make, and then some."

Expansion plans continued at Ford, despite the still-pending lawsuit, which was now under appeal, with the Dodge brothers and other Ford Motor Company shareholders. Edsel announced a planned tractor factory in Hamilton, Ohio, putting American cities on alert that there would be need for more. The $2 million Hamilton factory, reported the *New York Times*, would be the first in a network of hydro-powered factories brought online across the country.

"Mr. Ford has always wanted to advance the agricultural conditions of the world so that they would keep pace with those of manufacturing and transportation," the *New York Times* detailed. "He believes that he has made a step toward solving the problem of making the life of the farmer one that combines advantages of both country and city and making him industrious and contented during all seasons of the year."

As Ford now saw it, the Fordson would finally separate farmers from their land, allowing them to tend to their fields for only a short time each year, freeing them to work in his network of factories the rest of the time.[17]

The capital outlay and overhead outlined by Edsel Ford, Velie predicted, "would be quite a handicap."

Harold Dineen and Theo Brown left what may have been the last National Tractor Demonstration, encouraged that a contract with Ford, on their terms, would come soon. More importantly, concluded Velie, "we saw enough at Salina to convince us that the Waterloo Boy was not going to get a black eye in 1919, from the Fordson nor any other tractor. We have the best tractor on the market now, and we will have until the end of 1919 any way—so far, so good."[18]

For the first time, after a strong showing and a week together, executives at John Deere began to see the Fordson as a competitor.

Deere's dealers had grown impatient, raising questions about tractor availability, rumored updates to the Waterloo Boy, and the company's overall position on tractor sales. Deere tried to clarify its position, noting that despite advances in power farming, its major operations remained focused on the sale of agricultural implements for use with horses. For those looking to purchase a tractor, the Waterloo Boy tractor was to be sold as the "best and most efficient tractor." For those continuing to operate horses, which was, at present, 90 percent of American farms, John Deere offered a full line.[19]

Admittedly, replacement of the horse looked more like an inevitability than ever before. "The horse is going out of commission so rapidly," wrote George Peek from Washington, DC, as he continued to mobilize American manufacturing for war production, "that I fear an over-production on horse-drawn implements and an under-production on tractor-drawn tools of all kinds."[20]

Not unlike most of the industry, Deere remained caught on both sides of the transition. A few weeks after the Salina demonstration, Deere's board gathered, as they did weekly, in Moline.

The impact of the war, taking place thousands of miles away, felt much closer to home that week as the first gasless Sunday went into effect in Moline, though to Moliners the war was never far away. Easily visible from Moline sat Rock Island, a hundred-year-old military arsenal nestled on an island in the Mississippi River. More than 100,000 M1903 rifles and 160,000 howitzer shells would be manufactured on site before the war's end.

Americans, though, saw the temporary inconvenience of rationing as their patriotic contribution to the war, which looked to be turning in favor of the Allied forces. In Europe, the Hundred Days Offensive had begun, with the Allies reclaiming territory lost earlier that spring on the Western Front. The Battle of Amiens, commenced on August 8, was a carefully executed, modern campaign employing five hundred tanks, directed by low-flying airplanes, and tens of thousands of troops on the ground. The German army suffered more than thirty thousand casualties, compared to fewer than ten thousand for the Allies. A new offensive was now underway.

Deere's board met to consider an order for 2,500 No. 40 plows from Henry Ford & Son, deciding unanimously—they talked until it was unanimous—that until Ford reversed his dealer policy, they would not accept the order. For now, Harold Dineen, manager of the Deere Plow Works, was to proceed in making only one hundred plows for "trial purposes."[21]

Time would yet tell whether Deere's Fordson strategy would become a defining, or a debilitating, moment.

14

"BETTER, CHEAPER"

By the fall of 1918, the world looked as if it might be returning to a more recognizable state as colder temperatures began to arrive across the Midwest. Gasless Sundays were lifted in late October, spurring a "continuous auto parade," with droves of people finding their way to city parks the following Sunday in Moline. And if "the sunshine and fresh air are good influenza prophylactics, then Moline was thoroughly protected yesterday against the disease."[1]

News of the German surrender in early November spread like prairie fire. Even the farm tractor was receiving its due attention for contributions to the victory. In addition to increased food production, track-type tractors, most prominently the seventy-five horsepower Holt-75 gasoline tractor, hauled heavy artillery across battlefields throughout the war, even inspiring the development of the armored tank, a critical piece to the final Allied offensive.[2]

One could not help but be optimistic for the future.

Wartime orders, which triggered a manufacturing boom for some, were delaying growth for others and, in the worst cases, shattering their future. Willard Velie's plan to erect a tractor factory in East Moline was canceled because of the demand for Liberty trucks and, beginning in 1917, for three-quarter-ton trucks for the United States Army, every one shipped to the Mexican border. By the end of 1918, machine gun carts and ammunition wagons had earned the Velie Motor Corporation wartime contracts valued at $10 million.

Despite a flood of surplus government vehicles becoming available to the American consumer post-war, automobile sales boomed, with most companies expecting "more business than they can possibly deliver this year." William Durant's General Motors was investing nearly $38 million in expansions. Willys-Overland scheduled 180,000 cars. Dodge and Studebaker planned for 150,000. Ford planned for more than one million Model Ts for 1919, to be sold by thousands of distributors. Cumulatively, the automobile industry's record year produced nearly two million automobiles, still well short of demand.[3]

Velie, too, was "preparing for a huge volume of business during the new year." Plans for his new tractor factory would not be resurrected.[4]

Small tractor manufacturers were on the verge of being squeezed out by recent government restrictions on production. The Priorities Division of the War Industries Board, which regulated the supply of iron and steel available for tractor production, restricted companies that sold more than fifty tractors the previous year to only 75 percent of their steel and iron consumption for the year beginning October 1, 1918. Manufacturers who made between ten and fifty tractors the previous year could build no more than fifty. Manufacturers that sold ten or fewer could not make more than ten, but those restrictions were now lifted.[5]

Some industry insiders thought that Ford's tractor, in particular, was an inadequate introduction to power farming and that its inexpensive price tag and singular design would stunt industry development. Author Philip C. Rose, a critic of Ford's tractor even before it became available, opined that "his machine will not stand up" and "that he will find in short order." But full-line equipment manufacturers fully understood Ford's growing threat.[6]

In the spring of 1919, rumors began to circulate that Ford was buying a controlling interest in John Deere. In fact, reported the *Farm Implement News* to millions of readers, the deal was nearly complete. Howard Railsback, Deere's young public relations director, issued a denial. With ongoing negotiations regarding distribution happening in private, the evidence was hard to refute after the blossoming relationship had been put on public display in Salina less than six months before.[7]

Even more telling was the recent announcement that shook the automobile industry: Henry Ford was leaving the Ford Motor Company to start over, on his own, without shareholders.

In December 1918, nearing the end of the long-litigated suit between him and the rest of the Ford Motor Company's shareholders, Ford unexpectedly submitted his resignation, signaling plans "to devote my time to building up other organizations with which I am connected." Reluctantly, and surely suspiciously, the board accepted, naming Edsel to the top operational post.

The Ford family escaped Detroit for southern California, learning about the court's final verdict while on vacation. Now, including interest, the shareholders were due just less than $20 million in past dividends. The Dodge brothers' share amounted to a little more than $2 million.[8]

Not surprisingly, Ford had not retreated west to accept the verdict, but to enact a new scheme that would rid him of his shareholders once and for all. Charles Sorensen, who had been with Ford nearly since his start, later said that the tractor-making firm of Henry Ford & Son was created for two purposes. "One was to make farm tractors; the other was by threat of bringing out a new car to intimidate remaining stockholders to sell their Ford Motor Company shares." The latter plan was now in motion.[9]

The *Los Angeles Examiner* broke the news of Ford's "Huge New Company to Build a Better, Cheaper Car" on March 5. "I believe that every family should have a car and it can be done," Ford offered. "It is Mr. Ford's belief that when any corporation or organization dealing in commodities consumed by the public ceases to serve the public, its usefulness is ended and it naturally ceases to exist," he told the *Los Angeles Times*. Ford Motor Company executives had nothing to say and knew nothing of the endeavor and what would come of it, despite Edsel now at its head. He did not know, saying plainly that "the portion of it that does not belong to me cannot be sold to me, that I know."[10]

In fact, that very thing had already been put in motion. Behind the scenes, Edsel and chief engineer William B. Mayo, with the help of several bankers, began to acquire options on the outstanding stock, which became more volatile as Ford talked about his new company.

John and Horace Dodge issued a statement amid rumors that General Motors, tire magnate Harvey Firestone, and others were buying Ford out. "There would be no attempt to keep either Mr. Ford or his son in the firm if they simply wished to retire, but Henry Ford is under contract to the Ford Motor Co. and he will not be allowed to leave the firm and start a competitive business," they fired back.[11] In all, 8,300 shares, valued at more than $100 million, were purchased, largely with borrowed money. The bank's negotiators said Ford "danced a jig all around the room" when the deal closed.[12]

Ford released a statement confirming there would be no new company and that the Highland Park plant would continue to build the Model T. The Fordson would move there as well. Before the year was out, Ford Motor Company would build nearly 750,000 cars, almost 40 percent of total industry output.

⁂

To those paying attention, the farm equipment industry had gone through a significant rite of passage during the Great War. The industry had seemingly matured, it seemed, as consolidations and product standards began to separate manufacturers. Another outdoor National Demonstration took place in Wichita, Kansas, during the summer of 1919, but to less acclaim. George Friend, president and sales manager of the American Tractor Corporation, thought the show was "absolutely a waste of money." Deere's Howard Railsback agreed. He considered the show an "ancient and threadbare expression," pointing to "the fact that nowadays it isn't necessary for a farmer to travel a hundred miles to see a tractor pull a plow." As long as manufacturers felt a need for large demonstrations, the farmer would always consider tractors experimental, he reasoned. "Otherwise, why should they deem it necessary to hold demonstrations . . . ?"[13]

The Kansas City Tractor Club offered a reward for the "discovery of any man, woman, or child who attended the Salina demonstration and did not learn that there was to be a National Tractor Show in Kansas City in 1919." The 100-foot-long sign at the entrance and exit of the Salina show, along with advertising at field headquarters and at the Hotel Lamar, seemed to do the job.[14]

More than three hundred exhibitors participated, nearly seventy of them showing 137 tractor models—thirty-one tractors were being shown for the first time. Forty-three companies brought implements, and nearly two hundred exhibited parts and accessories. The show included more than 100,000 square feet of indoor floor space built by the club.[15]

International Harvester's marketing manager put it perfectly: "The industry was well represented . . . but the farmers were missing."[16]

Paul R. Preston, advertising manager for the Rock Island Plow Co., was struck by the "passing of the freaks." Preston, a Yale graduate, enlisted in officer training school for heavy artillery at Camp Taylor, near Louisville, after the Salina show, but with the end of the war returned to his advertising job in Rock Island, Moline's neighbor to the west. "Most of the tractors on display here are thoroughly practical. Some of the designers two or three years ago actually did not know how much power it took to pull the various agricultural tools through the ground under the varying conditions. How, under the circumstances, could they expect to design a tractor that was right? But that evil has almost disappeared. Most engineers now realize that the tractor must be completely 'agriculturalized' before it can perform successfully."[17]

There was still plenty of competition, less fraud, but a growing debate about the value of a tractor versus a team of horses. It was clear to most, though, that despite the latest wave of new models, the field of viable manufacturers was narrowing.

The Horse Association of America ramped up its criticism of the "desperate efforts being made by tractor manufacturers to meet the competition of horses and mules." The local and national press and implement dealers "are all being urged to coax you, individually, into buying a tractor," the association claimed.[18]

To entice farmers that summer, the price of a Fordson was reduced from $885 to $750. International Harvester followed, countering with a $225 reduction in the price of the Titan 10-20, now available for $1,000. Avery cut its 12-25 tractor price by $270 and its model 8-16 from $925 to $700. John Deere offered a more modest $100 discount on the Waterloo Boy, which still sold for $1,150.[19]

A $225 price reduction on the Titan 10-20 is advertised at a farm equipment exhibit, 1919. Wisconsin Historical Society, WHS-46207.

The cuts were expected to be temporary, an adjustment to rid manufacturers of existing inventories in a rapidly changing market. In fact, prices remained steady for the remainder of the year.

That fall, Deere's Theo Brown, now accompanied by Floyd Todd, lunched with Ford and Sorensen in Dearborn for nearly two hours. After learning all they could about "all the information that we desired regarding the Fordson tractor," they "settled with Mr. Ford such minor matter as the League of Nations, the reasons why we got into the world's war, the result of the world's war, his litigation with the *Tribune*, his political controversy with Newberry, his friendship for Wilson, and his attitude towards the constitution of the United States and the Monroe Doctrine."[20]

That summer, Ford had endeared himself even further to the American people after the verdict of a three-year lawsuit with the *Chicago Tribune* was issued. Ford had charged the *Tribune* with libel after a June 1916 article in the paper accused the manufacturer, based on his anti-war rhetoric, as "an anarchistic enemy of the nation." The case finally went to trial after nearly three years, culminating in eight full days of questioning of Ford himself. The defense, led by the same attorney who had defended the Dodge brothers in their case against Ford, worked to discredit the manufacturer with his sixth-grade education, successfully painting him as a "man with a vision distorted and limited by his lack of information."

During cross-examination, Ford was unable to answer basic questions about American history and government taught in schools across the country, showing a lack of understanding, or even familiarity with the events he claimed to revere. Some in the press vilified him as an embarrassment and a fraud. "The mystery is finally dispelled," editorialized *The Nation*. "Henry Ford is a Yankee mechanic, pure and simple, quite uneducated, with a mind unable to bite into any proposition outside of his automobile and tractor business . . . He has achieved wealth but not greatness."[21]

In August, a jury upheld Ford's case, granting him a ceremonial 6 cents in damages. More strikingly, reports of his failed testimony generated an outpouring of support from Americans across the country, especially rural Americans.

Theo Brown, a successful mechanic in his own right, had already succumbed to Henry Ford's plainspoken and outwardly simple philosophies on life and business, examples of which were reinforced in conversation on his many trips to the Ford farms and factories.

Sorensen took Brown and Todd through the Fordson plant after lunch. Down the center of the entire plant, "perhaps 800 to 1,000 feet," was a conveyor "propelled at a speed which experience has demonstrated conveys materials into the plant as rapidly as they can be consumed," they later reported.

Raw materials entered the plant and across the main conveyor, with each man machining each piece until ready for installation. Then it was removed and installed into the tractor itself. "This progressive method of manufacture terminates at the far end of the plant, when the complete tractor is cranked by machinery and run out under its own power, where it goes into the paint shop for painting and from there, as soon as dry, onto a car for shipment. No storage is provided for complete machines, as they are shipped as fast as made."

They were awed by the size of the workforce, amazed at the scale of production in a plant that Ford was preparing to relocate and improve upon at the River Rouge. On his office blackboard, Sorensen recorded a tally of finished tractors each hour, day, and the year to that point. Three hundred tractors were being built on average each day, reaching a one-day record of 314 completed machines. More than 82,000 tractors had been built year-to-date, nearly triple the previous year. The following year would be even greater.

Ford reassured his friends from Deere that "frankly that the best plow that has been made for the Fordson tractor is the JOHN DEERE." He urged them to work on a new grain binder and a seed drill as well, but felt "discouraged, however, in doing business with Deere . . ." They did not "have the proper vision," he criticized. Their latest proposal to build a thousand plows was "ridiculous." They should be building 100,000.

> *"We told him our sales policy," Brown and Todd reported on their return to Moline.*
>
> *"He asked us how long we had been in business. We told him eighty years. He asked us how much real assets we had accumulated in that time; we told him about $55,000,000.00."*
>
> *"Well," he said, "eighty years of your sales policy has accumulated you $55,000,000.00; my sales policy makes more than that for me each year—draw your own conclusions."*

Ford's proposition was given thorough consideration in Moline. Sorensen and Ford were still "most interested in our plow. They both felt we had missed a big opportunity in not selling our plows to Fordson distributors." But the most stunning news of all: Ford would not force implement sales on his dealer network. "They have cut out distributors, starting next August, and will not recommend any implements. They like our plow and would be very much pleased to have us get into production."

Despite the "advantages for the sale of the plow in greatly increased quantities" by selling through Ford dealers, Deere communicated to its branches, Ford fully understood "what it means to Deere & Company in the long run . . ."[22]

"Mr. Ford has personally tried out and approved the No. 40 plow," Deere's sales branches were told in late in 1919, in preparation for sales opportunities the following spring. Ford's contract with the Oliver Chilled Plow Company was set to expire July 31, 1920, as Ford had told Theo Brown.

"Mr. Ford has undoubtedly learned by experience that he should not handicap the sale of this tractor by permitting the distributors, who are not implement men, to insist that the purchasers of Fordson tractors buy an implement from them . . ." Deere reassured its representatives that "when a dealer places his confidence in Deere & Company and links his fortune with us and agrees to handle the John Deere line, that we have a great obligation and responsibility in seeing that he obtains the implements which he needs to properly take care of his trade."

To back up its claim, nearly $500,000 was allocated for increased production of both tractors and engines in Waterloo.

15

"DEPRESSION IS AWFUL"

To date, tractors had been designed from the collective knowledge of engineering principles, with cues from automotive manufacturers, not from actual crop practices. *Implement & Tractor Trade Journal's* wrap-up of the 1919 tractor show in Wichita pointed out that ". . . one of the great troubles is that the sales departments too often determine the policies pursued by given companies, with the result that the quality consideration is invariably and inevitably overruled by the sales consideration." By now, it was being argued, many of the tractors looked the same, and that to the naked eye little difference could be found. Now, tractor design was coming under increased scrutiny as buyers became better educated. Price was still a key factor, but an understanding of the daily demands of operation was creating new expectations. Reliable starting, steady engine performance, a hardened crankcase, the protection of parts from dirt, and prevention from overheating were just a few of the growing list of demands from farmers.[1]

Versatility was becoming increasingly important. In Blue Mound, Illinois, the first corn belt cultivator demonstration was held over two days in early June

1919. The event featured motor cultivators from Avery, Emerson-Brantingham, the Moline Plow Company, Toro Motor Company, Allis-Chalmers, and International Harvester. The motorized cultivator was earning a strong following.

The demonstration in Wichita a few months later featured those six machines, plus two more, each looking to add to the Moline Universal's recent success. A recent infusion of capital from automaker John Willys propelled the company to sales of six thousand machines, or 4 percent of the total tractor market in 1919.

Willys had saved the Overland Automotive Division of the Standard Wheel Company in 1908, a company a day away from receivership, and transformed it into the second-largest automobile manufacturer in the country, ahead of General Motors and behind only the Ford Motor Company (he renamed it the Willys-Overland Motor Company in 1912). Growth came through the constant acquisition of companies, brands, and, through them, dealerships. In 1918, the Moline Plow Company was added to his portfolio of brands.[2]

Wartime restrictions on steel and other raw materials incentivized automakers to enter the tractor business, giving access to additional materials with the addition of the new product line. In 1917, General Motors acquired the Samson Tractor Works, the "tractor that the General Motors Co., a $170,000,000 concern, decided was the best in America," explained H.A. McMullin & Son, Samson tractor distributors in New Mexico.[3]

General Motors's William Durant poured money into an expanded tractor factory in Janesville, Wisconsin, which for two decades had been turning out plows, cultivators, and other implements and was now the company's center of operations for the manufacture of tractors and trucks. He also bought the rights to the Jim Dandy Motor Cultivator, a rein steer tractor guided by leather reins instead of a steering wheel.

The 1918 Samson Sieve models General Motors brought to the National Tractor Demonstration in Salina were a pricy $1,750. A more competitive model based on the Fordson, the Model M, was ready in May 1919, just as Durant was forced out of General Motors. The M sold for a more competitive $650. The price included fenders, belt pulley, and a rear platform to allow for operation while standing—all extra with the purchase of a Fordson. Even the color on the Fordson was matched.

Over the next few years, General Motors would invest more than $20 million into tractor operations—$3 million in "overrun expenditures" alone. Annual capacity at Janesville could reportedly accommodate 120,000 tractors.[4]

William Durant was on the right track, as manufacturers and entrepreneurs looked harder at the tractor form that coupled the characteristics of a horse with the power of a tractor. The term "general purpose" continued to appear in articles and discussions about the tractor's ultimate ability do more around the farm, especially if it were truly to offer replacement value for a team of horses. Yet "no development in the industry was regarded with more distrust and wholesale opposition than the suggested general-purpose tractor," later wrote G.D. Jones in an article for *Agricultural Engineering*. Most of that opposition came from inside the companies designing them, he offered.[5]

The persistent question of power farming was not the quality of farm work, but instead a purely economic issue, whether tractors could cover the up-front investment, coupled with the purchase of new implements designed for tractor use, and still deliver a profit. A manpower shortage before 1914 helped the tractor's cause and now had become a post-war crisis as returning soldiers, continuing the pre-war trend, flocked to the city instead of returning to the farm. Manufacturers of new technology, from radios, stationary engines, tractors, trucks, and more, attempted to make the farm more appealing to young people. A new Indian Big Twin motorcycle could "kill the spirit of discontent that is luring him from the soil to the bright lights of the cities," claimed one.[6]

Fewer hands continued to feed demand, but the needs of each farm were different. Farmers raising hay and small grain could more successfully replace horses with tractors, but it was not the case in the Midwest, where corn was king. As one author put it, a "corn field is an arbitrary proposition."

More reputable data was being published weekly. After testing at a hundred farms across the state of Illinois, the Agricultural Experiment Station of Illinois reported an average plowing time of 16.8 days, at an average of seven acres a day. With a three-plow tractor, a farmer could accomplish 50 to 75 percent more work than with the usual four- or five-horse team. Typically left out of the discussion was that the farmer still needed the horses for other operations, so the tractor was still an additional, not a replacement, cost.[7]

New scientific methods, combined with education on progressive, or "modern," farming techniques, infiltrated the daily life of farmers like never before. Where previously there was a dearth of information, farmers of the early twentieth century had streams of scientific data available to them. The Department of Agriculture published bulletins. National, regional, and local organizations published their own educational literature. Manufacturers began to offer their own experts as well. Local extension offices studied soil composition, crop rotation, equipment efficiencies, air temperatures, rain, drought, soil compaction, irrigation, and dozens of other data points that farmers were only beginning to scientifically understand. They also fought new versions of age-old problems, pests and disease chief among them. More than two hundred pests had been identified as threats to corn alone. It was an entirely new type of trench warfare.

The chinch bug, which wintered in field trash or high grass along fence lines, deposited eggs at the base of young wheat in the spring. Hatching in the summer, they fed on the green vegetation, then moved onto the corn fields, "sucking the life-giving juices from the plant, destroying the entire field in a few days."

The European corn borer arrived in the United States between 1910 and 1918, coming from infested broom corn landing in Massachusetts and New York. By 1917, the female moth, capable of laying hundreds of eggs, had infected much of New England, then spread north to Ontario and then into the Corn Belt.

The borer did not stop with corn and moved to celery, potatoes, tomato vines, beets, and rhubarb. W.R. Walton, a government entomologist, recorded "no less" than 167 plant varieties so far impacted.[8]

Seventy-five percent of the world's corn was grown in the United States, which would produce three billion bushels in 1923, a slight decrease from prewar, but well above the two billion–bushel output of 1891. Hundreds of varieties were being grown, the planting window beginning at Mother Nature's discretion when the leaf buds began to unfold on the trees. Breeding had created dozens of corn types and varieties, known by names such as Silver Mine, Golden Eagle, Legal Tender, Flint, and Boone County White. In central Georgia, central and southern South Carolina up through central Arkansas, planting occurred as early as mid-March. In parts of Florida, Texas, and Louisiana, planting could begin as early as February. Working north, planting in northern Wisconsin, central Michigan, and central New York occurred as late as mid-May.[9]

Corn was the most important livestock feed and was also used for corn oil, hominy, corn flakes, corn bran, corn syrup, corn starch for food and laundry, and, of course, liquor. Corn cobs were used for fuel, fertilizer, and feed, and to make tobacco pipes. The pith of the corn stalk was used as a lining for battleships, and cellulose from the stalk in the manufacture of paper and gun-cotton. Husks were used to make horse collars and door mats. No part was left to waste.

Corn accounted for nearly $4 billion in national crop value. It was also a source of national pride. Corn husking competitions drew tens of thousands of spectators to witness the country's fast picker, a path to instant celebrity. Deere's expert, Dr. Warren E. Taylor, calculated that the corn yield in Iowa and Illinois, the "garden of the world," was averaging thirty-two to thirty-five bushels per acre in the early 1920s and was "not enough; it is not one-half or one-third as much as every acre of that rich territory should produce." The Corn Belt now included the Dakotas, Minnesota, and Wisconsin.[10]

In the field, corn was planted in rows spaced anywhere from three to six feet apart, and anywhere from eight to twelve inches deep, depending on soil type. These factors and many more determined the type of planter, drill, cultivator, and a long list of other machines essential to planting, cultivating, and harvesting the crop. No tractor had yet been designed narrow enough to straddle the rows, nor high enough to cultivate without damaging the emerging stalks. The market potential was even larger than the existing market served by the Fordson and the hundreds of competitive tractors available. These tractors plowed—pulling one, two, three, or more. Some ran threshing machines from their belt pulleys and provided power for other odd jobs like pulling stumps, hauling trailers, or even pushing snow. But they lacked versatility on the farm.

❧ ❧ ❧

There were signs. But when faced with unparalleled demand and record profits, tractor manufacturers preferred to embrace their newfound prosperity. Postwar America was a land of mobility and freedom for sure, but a growing disparity was further evident as a smaller percentage of the population controlled more of the nation's wealth. A world war, a global pandemic that would claim the lives of nearly 700,000 Americans, and a growing reliance on credit for financing new and larger purchases hinted at darker days ahead.

Tractor-makers in particular had become convinced by their own advertising, sure that they had turned the corner and won the argument that power farming was indeed the future. In reality, annual growth had been buoyed by a strong export market in the final few years of the war to Great Britain, Canada, and beyond. Further incremental gains were realized, mostly by Ford and International Harvester, which combined to build half of the tractors in the United States in 1919.

According to the Food Administration, the United States shipped more than three billion pounds of meat, dairy, and vegetable oils to the Allies, a $1.4 billion increase in 1918 compared to the previous year. The estimated wheat yield in the United States reached nearly one billion bushels.[11]

With peace came further readjustment. After the planting of the winter wheat crop in the fall of 1919, announcement came of changes to the government's guaranteed purchase price, which had risen to $2.26 a bushel. The change would take effect on July 1, 1920.

"Depression moves from the East to the West," International Harvester president Cyrus McCormick, Jr., later wrote, noting that "scattered farmers of the country feel trends and tendencies less quickly than do the organized inhabitants of an urban community."

The farm equipment business was cyclical and volatile, a "long turnover business." Those were the only guarantees from year to year. Most farm equipment was put to work for only a short time each year during small planting and harvesting windows. In some cases, such as the case with a binder or a hay loader, the implement might be used for only four or five days in a year. Which days depended entirely on the weather.

A majority of purchases were made near the first time of use, making sales forecasts tricky and unreliable. Yet, to meet demand, raw materials were ordered a year in advance, implements were built, and, if left unsold, could remain in inventory for a year or more. Ongoing freight and storage costs weighed heavily on manufacturers. Territory canvassing by salesmen in the fall drove orders for the spring—dealer, sales branch, and factory all hoped the orders held up when it was time to deliver.

In January 1920, Leon Clausen prepared Deere executives for the large financial investments required for the 1921 selling season due to the "increasing importance of taking care of our dealers . . ." It would take a year at minimum, he told them, to properly plan and install equipment and train the staff at the Waterloo tractor and engine factory "to turn out good work." Deere should be prepared to build 7,500 tractors in 1921, estimating a market for one million to 1.25 million farmers in the country that will use tractors on their farm. "I believe there is a business of something like 200,000 tractors per annum," a number that proved wholly accurate by the end of the year. After that, a replacement business of 10 to 15 percent annually would put industry production at as many as 175,000 tractors. In this, all manufacturers would find, the numbers were optimistic.

"Deere & Company certainly is entitled to a fair share of this business," Clausen stressed. If they couldn't capture at least 10 percent of industry sales, "there will be something wrong about our machines or about our methods of doing business." On his recommendation, nearly $400,000 in additional equipment was allocated for the Waterloo tractor and engine factory.

Just in case prices failed to hold, plans for an updated Waterloo Boy were moved up, expediting a change in its color scheme to Deere's standard green and yellow, a slight modification from the Waterloo Boy's previous colors. A new decal was added to the front frame on each side, a leaping deer, trademarked by Deere in 1876, that appeared to jump from the white, cloud-like background.[12]

Additionally, "immediate plans" were enacted to build at least forty tractors a day.

By mid-May 1920, the Waterloo Works was still operating "approximately on schedule." An updated Waterloo Boy with a redesigned kerosene-burning engine was being field tested and reported as "very favorable."[13]

❦ ❦ ❦

The Fordson ended its run at Highland Park at the end of September 1920. Production of an updated Fordson Model F tractor, with subtle cosmetic changes such as seven-spoke wheels and plain radiator sides, began again five months later in Building B of the Ford Motor Company's new River Rouge factory. The Hercules engine was replaced by one of Ford's own design, but more or less, the Fordson was unchanged. Most importantly, it remained inexpensive.[14]

Ford again confirmed plans to sell a million tractors a year. Few could have imagined it even ten years before, but there were now more than 300,000 tractors on American farms. The number of farms with tractors had doubled since 1918 alone, yet still less than 6 percent of farms nationwide had one in operation.

Through the tractor, Ford predicted, farmers would only actually need to work a few weeks a year and could "employ themselves profitably for the 200 days each year while there is no work to be done on their farms."

Fordson tractors would soon multiply in numbers comparable to the Model T, and the Ford and Fordson names would in short time be on every shed, garage, field, and street across America.

Despite the emergence of hundreds of tractor-makers, the mergers and consolidations of farm equipment manufacturers had brought increased stability to the market since the turn of the century, even through the years of war. Implement sales more than doubled during the course of the war to $213 million. Sales increased to nearly $300 million by 1921. Now, new pressures followed. For the remaining manufacturers, the cost of tractor research and development, design and introduction of compatible implements, and new advertising skyrocketed.[15]

Harvester's hard-nosed general manager, Alexander Legge, resolute as ever, told an audience of Chicago businessmen that farmers were suffering from the same problems as all businessmen—"efficient management, production, and

Henry and Edsel Ford with their namesake, the Fordson tractor, 1921. Image from the Collections of The Henry Ford.

marketing." Farmers were working harder, investing more, but earning less. No wonder young people continued to leave the farm. There were more than 1.5 million fewer people living on farms in 1920 as compared to 1910.[16]

Despite favorable weather, increased mechanization, and growing export markets, farmer incomes, on average, were seeing little gain from the post-war boom. Yet, somehow they were still buying cars.

In 1921 and 1922, the value of Ford Motor Company automobile sales to farmers exceeded farm machinery sales by all manufacturers combined. It was no surprise. Half of the nation's population was classified as farmers, or at least rural residents, and Ford held half of the automobile market.

Deere executives were perplexed by the "inexplicable" sale of automobiles to farmers. "Ford's business, as we all know, continues highly profitable, and Ford sells most of his cars to farmers. It is the farmer trade that has made Henry one of the richest men in the world." Meanwhile, farmers continued to defer the purchase of new implements, putting their money instead into an upgraded automobile. But the time was coming, offered one optimistic writer, when farmers would choose new implements over a new car.[17]

"When you get to making the cars in quantity, you can make them cheaper, and when you make them cheaper you can get more people with enough money to buy them," Ford said repeatedly in one form or another. "The market will take care of itself . . ."[18]

Ford had spent more than $100 million over the previous three years on construction of the River Rouge factory, dividend payments, and the purchase of timber fields and mines to support the growth of the Ford Motor Company. To drive continued sales, he slashed prices.

His executive team attempted to fight the cuts, but stood little chance. The price of a runabout was reduced from $550 to $395, a sedan from $975 to $795. "Inflated prices always retard progress," Ford offered in a statement. "We had to stand it during the war, although it wasn't right, so the Ford Motor Company will make the prices of its products the same as they were before the war."

Ford's willingness to lose money to achieve sales confounded those around him, including his son Edsel, who fruitlessly argued against the tactic. In fact, Ford was losing $20 on the sale of every car, working to squeeze parts suppliers and dealers to cover the gap. Competitors had no choice but to follow suit.[19]

❧ ❧ ❧

In the first half of 1921, most of John Deere's factories reduced production schedules, then began to temporarily shut down completely. The "economy-use of electricity" at the corporate office in downtown Moline was stressed that spring. Lights and use of the elevator cost an average of $25 a day. Not since Moliners contributed to the war through the rationing of gas and heating oil with lightless nights and gasless Sundays did things feel so bleak.[20]

Car sales slumped at the Velie Motors Corporation, and in April, Willard Velie finally withdrew completely from the board of directors. "My personal and more important interests under existing conditions will require of me all the thought and effort that I care to expend," he wrote in his resignation letter. It would nearly bankrupt him.[21]

The Oliver Chilled Plow Company, one of Deere's most formidable plow competitors, increased its discount from 8 percent to 20 percent. "The reason they did this is apparent to everyone," Deere sales branches were told. They left horse farmers behind and banked their future on tractor plows for the Fordson. "They now find the tractor manufacturers and tractor distributors, with their factories and organizations closed down or running on a greatly reduced schedule, and they figure that something must be done to buy back the good will of the dealers which they so flippantly threw away."[22]

Theo Brown returned to Dearborn for the first time in nearly a year. With the distribution of implements for the Fordson settled, there were only fine details left to discuss on the introduction of the No. 40 Plow. Henry Ford further offered to "do everything he could to help us," but Charles Sorensen, after Ford's departure, shared he was "sore" that Deere had armed its distributors with a facsimile of the letter from Edsel Ford endorsing the Deere plow. He "thought it was a breach of confidence to do so."

Later that summer Edsel wrote directly, requesting that Deere pull its advertising campaign, which featured the text of Henry Ford's personal approval. "The reason for this is that we are informed the Ford people feel that in making further general use of the letter we are simply trying to use Ford as an advertising medium," the sales branches were told. "Please understand, also," the bulletin closed, "that the friendly relations existing between this company and the Ford Motor Company continue."[23]

❧ ❧ ❧

"The business depression is awful," wrote Theo Brown in his journal on August 30, 1921. "The shops have all been shut for a long time and there is no prospect of any immediate improvement." Consolidated branch sales for the first three quarters were down 63 percent.[24]

At year's end, Deere addressed the slumping farm economy in its annual report to shareholders. "The unprecedented loss from operations was due to the great depression in the price of farm products, and the consequent impairment of the buying power of the farmer, who is the only user of our product." Deere

& Company reported a $2.7 million loss in 1921, paid a full dividend of more than $5 million out of the surplus account, and reported inventory losses of nearly $1 million. Fortunately, Deere held nearly $10 million in cash reserves, cushioning the blow and keeping the company on solid financial footing.

The near term still looked bleak, though. Factories were closed for the fall and into the spring of 1922. In 1922, the surplus fell further, and sales at 40 percent of normal volume were reported for the year.[25]

Deere converted $10 million of debt into ten-year gold notes "owing to the business conditions already referred to, and to the fact that for some few years, at least, the farmers of the country will require longer credits," Butterworth told the *New York Times*.[26]

Tractor build schedules plummeted from forty per day in early 1921 down to twenty by September, and inventories were accumulating. Waterloo Boy model N sales fell from 5,045 tractors in 1920 to seventy-nine in 1921. Three years into the acquisition of the Waterloo Gasoline Engine Company, Deere had yet to realize a profit. Nor were there now prospects for one.[27]

The depression had heightened consequences for implement manufacturers as product piled up in warehouses across the country. Deere had several thousand tractor plows sitting in warehouses for more than a year. "We already had suffered a large loss upon them," wrote Deere's Clausen, "but within the past thirty days they have been rendered almost valueless by the action of the International Harvester Company."[28]

Harvester pumped an extra $100,000 into advertising but to little result. By the end of 1920, they had accepted more than $41 million in canceled contracts on 312,000 machines.[29]

Like its competitors, Harvester suffered steady domestic sales during the war, but even profitable war contracts were marginalized by the high cost of hard-to-secure raw materials. The dismantling of its harvester business and the separation of its two-brand dealer network by the Federal Trade Commission, combined with a unique dependence on international sales—upward of 40 percent of sales coming from outside of North America—exposed Harvester on all fronts, making the company visibly vulnerable for the first time.

16

FARMALL

The timing could not have been worse when on December 13, 1920, moving pictures of International Harvester's new experimental tractor were shown to executives in room 517 of the Harvester Building, Chicago. Present were chairman Cyrus McCormick, Jr., acting president Harold McCormick, general manager Alexander Legge, and others. There was a leadership transition underway as Cyrus transitioned the presidency to his brother, Harold, who preferred not to have the responsibility. Behind them, a reel of film was removed from a noisy tin can and placed expertly onto the projector, a descendant of Edison's kinetoscope. In front of them, projected on the screen from the small lens of what had the appearance of a microscope lying on its back, was a new experimental tractor trailed by a ten-foot binder and shocker. One man operated the entire outfit. It did the work of three men and a team of eight horses.

Disappointingly, at least to the engineers in the room, "there did not seem to be much enthusiasm with reference to the experimental tractor," wrote one insider. Monies were approved to build five more tractors. A month later, the number was reduced to two. Any enthusiasm for the tractor and its possibilities was tempered by an uncertain future in the agricultural equipment business.[1]

Harvester designers had continued work on their general-purpose tractor concept purely out of conviction, but no official developmental program was in place. Bert Benjamin, now head of the experimental department at the McCormick Works, and others used scrap parts to assemble new forms and at times even test machines on the Tractor Works' test track.

The resistance from leadership was understandable. Several versions of their motor cultivator, an early general-purpose-type tractor, had failed to find an audience. One hundred machines were built in 1917, but only half of them ever left the factory. Another three hundred were built in 1918, making appearances in Salina and other field demonstrations. Half of those were sold in 1918, the rest carrying over into 1919, 1920, and 1921 before the inventory was finally depleted. It was a rare commercial failure, partly due to high manufacturing costs. But its designers remained convinced that the idea of an "all-purpose" tractor, one that replaced horses entirely, had merit, and continued to experiment with the form. Importantly, work on a corn planter, front mower, sweep rake, and binder attachment had all progressed.[2]

Early attention was paid to solving rollovers, a common problem in lightweight tractors when working uneven surfaces, or tipping backward from the force of a heavy drawbar load. Both the Bull and the Fordson were notorious

The International Harvester motor cultivator. Courtesy John Deere.

for such accidents. Efforts to increase the wheel weight, to outfit the machine with crawler tracks, effectively eliminating its use as a plowing tractor, were employed by Harvester but still failed to solve the issue.

"When we were building attachments to fit a tractor we were not getting anywhere in particular," it was noted later, "but when we started building a tractor to meet the requirements of the attachments, we began to forge ahead."[3]

A "radical change" was made early in 1920. The two front wheels were converted into traction wheels, using a differential to transmit engine power to the wheels with the ability to rotate at different speeds. The following year, they flipped the running gear, putting the large traction wheels in back and a single, smaller wheel in front. The engine was moved from the rear to a position in the middle, and parallel to the frame.

Cultivating equipment was mounted to the front, and three reverse speeds were added. The seat could be reversed so the machine could be driven in two directions, similar to motor plow designs like the much-copied Hackney, and even Deere's first experimental Melvin tractor in 1912.

After the lukewarm reception from Harvester management, modest resources were allocated toward development of the Farmall through three separately managed programs in 1921, each independent of the other, but with a singular vision: (1) development of the Farmall tractor at the Chicago Tractor Works, (2) development of Farmall attachments by the experimental department at the McCormick Works, and (3) development of the new 8-16 and 12-25 tractors for 1922. This work would be done at the Deering Works, seventy-six acres of facilities for the production of corn binders, corn pickers, harvester-threshers, grain headers, rakes, potato diggers, and culti-packers. An ambitious goal, which was soon reduced, was set for one hundred Farmall tractors, all hand built, for 1922. They would, "naturally, cost far above what the ultimate manufacturing cost would be if they were put into production."

In December 1921, management approved building twenty Farmall tractors, twenty cultivating attachments, twenty mowers, twenty corn planters, and ten binders. Further experimental work on a shock gatherer, corn picker, side delivery rake, and other attachments for the Farmall would continue as well.

A breakthrough was soon made on the cultivator, which was built to line up with the front wheels and give it a greater amount of side travel to respond to a crooked row.

The engineering teams considered their light tractor, and the cultivator solution, a "turning point." They replaced the chain drive with a bull-wheel drive, redesigned the power lift, an ingenious mechanism to lower and raise the implement from the seat, and moved the belt pulley under the seat.

Bert Benjamin and his team of engineers were thrilled. The sales force, burdened with the failure of Harvester's motor cultivator program, was not. Nor was Alexander Legge, resolved to fight Henry Ford's price cuts to the bitter end.

17

IRON MAN

Harold McCormick resigned as president of the International Harvester Company in the summer of 1922, not long after inheriting the position from his older brother, Cyrus McCormick, Jr. Harold's personal affairs, which received far greater attention than his business acumen, were taking up much of his time. He had only recently divorced to marry his second of three wives, a Polish opera singer named Ganna Walksa. It was her fourth marriage.

Well prior to McCormick's resignation, it was clear to all who was in charge. When Alexander Legge, now a twenty-six-year veteran of the company, was officially named president of what a Deere memorandum called "head of the vast and far-flung company," it was merely a formality.[1]

Legge stood six-feet, two-inches tall and was "built proportionately." He was a long-time force at International Harvester, and even more so on the world stage. A "tall, awkward, unpretentious man," Legge was a former cowboy, mechanic, and, at Harvester, a legendary salesman. His friends simply called him Alex.[2]

Alexander Legge moving straw. A McCormick-Deering tractor operates in the background, 1923. Wisconsin Historical Society, WHS-11477.

Legge was born on a Dane County, Wisconsin, farm a year after the end of the Civil War. His parents, Scottish immigrants, lost the farm in 1875 after years of mounting crop failures. His father bought into a partnership the following year, on the invitation of a family friend, at a new 2,000-acre ranch in Colfax County, Nebraska. Young Alex grew up on the ranch. At age fifteen he was hired on as a temporary foreman when his father was away and was soon being sent on trips to buy cattle. In February 1883, he barely survived a blizzard, his horse leading him home from school through temperatures that hit 30 degrees below zero. Both Legge and his best friend, Charlie Wertz, whom Legge's mother had sent out to find him, returned nearly dead. The episode caused a pulmonary condition that troubled him throughout his life. The doctor recommended a drier climate and a move west.

In 1887, Legge and Wertz were hired at the V-R Ranch outside of Douglas, Wyoming, a twelve-thousand-head cattle ranch that also raised horses for the United States Cavalry. The two were hired as "fence riders," protecting the herd for $35 a month. Both were promoted to payroll courier, carrying at times thousands of dollars in cash for many of the local businesses that used the V-R Ranch as a centralized supply station.

At the age of twenty-one Legge returned to Nebraska to help care for his ill mother. She recovered, and he stayed to again help his father, this time on the 240-acre farm he now operated. It was on his father's farm that Alex purchased his first threshing machine. From it, he learned of his strong mechanical ability, finding temporary work with a local dealer who also asked him to help collecting debts for the McCormick Harvesting Machine Company in Omaha. He soon earned a full-time position collecting debts and repossessing equipment.

Legge once visited a farmer who had offered up a bull as collateral. When he asked to see the bull, the farmer pointed to the field, told him it was dead and buried, and that there was neither bull nor money to pay the claim. Legge asked for a shovel, planning to at least take the bones to a rendering plant for partial payment. The farmer settled the account.

His rise at Harvester was swift, and by the turn of the century he was running McCormick's Omaha branch. Soon after, he was in Chicago organizing the collection system for the McCormick Company worldwide; then, as assistant general manager of sales, he developed collection systems for the newly formed International Harvester Company.

In the new International Harvester Company, Legge spent time in Western Europe and pre-revolution Russia building Harvester's foreign operations. In July 1917, he joined Deere's George Peek at the War Industries Board, leveraging American manufacturing to streamline military production. Bernard Baruch, head of the Raw Materials Division, headed a nationwide search for a "right-hand man," and Legge's name turned up more than any other. Leland Summers, a board advisor, told Baruch, "There's your man! He knows Europe, knows human nature, is a shrewd trader, as straight as a die and an unbeatable fighter. His is the best mind in the International Harvester Company—but I don't think you can get him.'"[3]

Legge was appointed vice chairman. Post-war, Baruch asked Legge to work on the economic terms of the Treaty of Versailles. "His gift of hard common sense was nothing short of genius," wrote Baruch. "The treaty-makers on the other side were wizards in diplomacy and economics, but Alex was fully capable of meeting them. His mental shortcuts were amazing and his European business experience had given him a specific and comprehensive knowledge of the value of things over there. This was invaluable in connection with reparations estimates."[4]

Legge's work as primary reparations negotiator at the Treaty of Versailles would soon be recognized with the Distinguished Service Medal. Over time, his war service would be recognized with the Commander of the Crown of Belgium, Officer of the Legion of Honor of France, and Officer Knight of the Crown of Italy.[5]

In early February 1922, Legge and Cyrus McCormick, Jr., who remained chairman of the board, visited their new motor-truck installation at the Springfield Works. In addition to new tractor development already underway, Legge was working to revitalize the company's floundering truck business.

The Harvester tractor lineup now featured the two-cylinder Titan 10-20, the International 8-16, which would outsell the 10-20 for the first time in 1922, and the larger International 15-30. The Titan 10-20, first introduced in 1916, was now being offered on an installment plan, an effort to extend its life a bit longer. "The new plan is temporary," customers were told, "to relieve the present unusual financial situation, and is therefore subject to withdrawal at any time." Furthermore, if the price was decreased in the near future, customers would be refunded the difference.[6]

Development of the International 15-30 began well before Ford's first price cut triggered a landslide of price reductions, and despite depressed prices and low sales expectations, the work continued. The new tractor was designed to pull three plows, same as the Waterloo Boy, so "owners can do their work easily and in comfort." Nearly two hundred 15-30s were put into service by the end of 1921, achieving sales of more than 1,500 machines in 1922.[7]

The introduction of the "time payment plan" for the Titan 10-20 was essential to clearing inventory of the older, smaller tractor. A 10-20 version of the new 15-30 was authorized late in 1921. It would replace the time-tested Titan.

During the inspection, McCormick and Legge were notified of an urgent phone call from Chicago. Ushered into the office, Legge grabbed the receiver and put it to his ear. McCormick watched and listened; it was all he could do, while there was "much talk on the other end, and then an explosion from Alex." At the beginning of the year, the Fordson tractor sold for $790 "stripped." With a governor (a device that controlled engine speed), fenders, belt pulley, and accounting for shipping, the cost rose to over $950. In the two months since, the price had fallen to $625. Now, Ford cut the price of a Fordson further, to $395 "stripped."

"What'll we do about it?" Legge screamed rhetorically into the receiver. "Why, damn it all—meet him, of course! We're going to stay in the tractor business."

In the upcoming months, dealers were directed to offer a free plow with the purchase of a Titan, which was being sold at an already reduced price. If they heard of a Fordson demonstration, they were to challenge it head-to-head in a field trial. "The International Harvester Company yesterday initiated a price war with Mr. Ford," reported the *Chicago Journal of Commerce*, noting that the "new prices quoted on tractors and the temper of Harvester officials would seem to indicate that the prayers of the country that someone give Mr. Ford some real competition are about to be answered." Harvester rationalized in an official company statement that the "reduction is not justified by any present or prospective reduction of manufacturing costs," but was made purely to keep dealers competitive.

The Horse Association of America added to International Harvester's response, with additional arguments against the Fordson and questioning whether farmers even needed a tractor. "Can you take it apart and make two, three, and four separate units out of it? Can you buy a tractor, use it three years, and sell it for more than it cost?" A statistical analysis of tractor farming, the association concluded, demonstrated that there was no field work that could not be done more cheaply and efficiently with horse and mules. "THEREFORE—there is no place in sound farm management for either."[8]

⁂

Like the Model T, the Fordson was losing money on every sale as well. Edsel Ford had already admitted, before the cuts, a loss of $100 per tractor. A $230 reduction, to $395, only compounded the loss. The *Chicago Journal of Commerce* easily surmised that the reduction indicated Henry Ford's willingness to personally absorb the losses to achieve the targeted annual production of one million tractors. Farmers interpreted it as Ford incurring a loss on their behalf. In reality, Ford was pushing the new losses on "protesting dealers" who were already operating on slim margins for tractor sales.

For now, the Fordson and its competitors "will be compared in every corner grocery store in the country from today until the price war is over."[9]

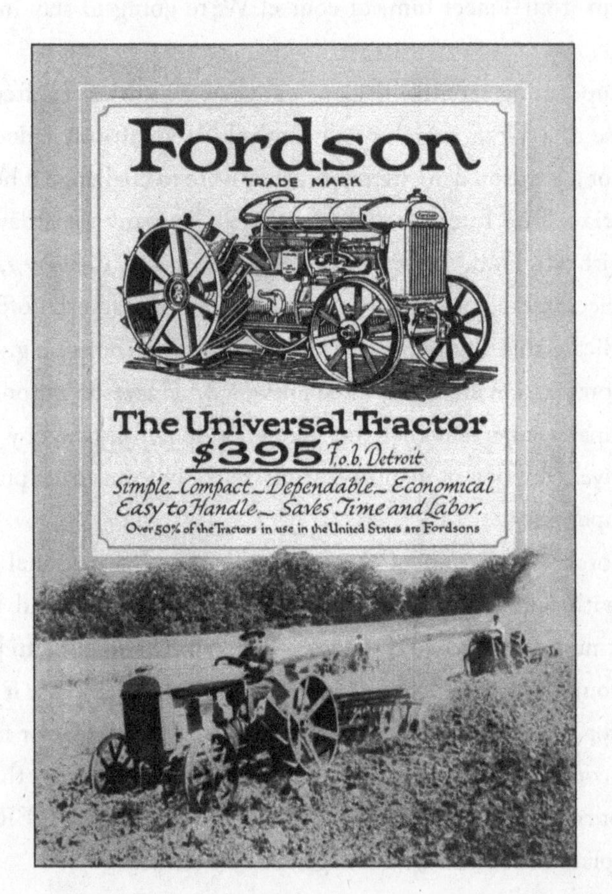

Henry Ford's answer to industry overproduction was the $395 Fordson. Wisconsin Historical Society, WHS-96641.

A factory representative confirmed that the $395 price was still in effect for the Fordson. He knew nothing of future reductions, "but one can never tell." A representative from the sales department said capacity was now at 250,000 tractors, far from the million Ford was promising, but far exceeding what the entire industry had ever sold in a year. Actual production, though the number was not shared publicly, would reach only a quarter of capacity that year.[10]

"It is important that it shall be cheap," Henry Ford offered. "Otherwise power will not go to all the farms. And they must all of them have power." In a few years, a farm using horses will be "as much of a curiosity as a factory run by a treadmill."[11]

Ford distributors continued to show the Fordson at tractor schools and local demonstrations, selling just as much the virtues of Ford himself, who was just like them, imparting "the fact that Mr. Ford himself is a real farmer and that his highest aim in life is to help the farmers of the world to lighten their burdens and make farm life more pleasant." With the price drop, every farmer should "take advantage of the opportunity Mr. Ford has given them."[12]

18

"OUR MAIN COMPETITION"

John Deere sold an anemic seventy-nine Waterloo Boys in 1921, the tractor line accounting for 3 percent of total sales. With dim prospects for 1922, the price was cut again, a total reduction of $750 from its original price. Nearly three thousand already-built tractors remained in inventory. Deere, like Harvester, refused to push them onto dealers.

Deere's Waterloo operation was on its way to a nearly $500,000 loss for the year—a significant improvement over the previous year.

Deere's board of directors pushed for the design of a harvester for the Fordson, despite the insistence of Willard Morgan, manager of its Harvester Works, that it was "not practical commercially" to design a machine just for the Fordson.

Floyd Todd, as the result of "developments in the tractor field," pushed for an updated stationary engine, "as there is an extensive market for this product, which will bring us some return on our investment in Waterloo."

The tractor program was under increased scrutiny, all knowing that "we might be faced with the problem of discontinuing the manufacture of tractors

Waterloo Boy tractors outside of the Waterloo Gasoline Engine Company factory, circa 1920. Courtesy John Deere.

at the Waterloo Plant at some later date . . ." A proposal to spend $4,000 to advertise a price cut was voted down. It would only compound the losses.[1]

In April 1922, Leon Clausen completed a study of potential reductions in Waterloo. Prophetically, though not to anyone in the room at the time, several key decisions would determine the course of Deere tractor production for the next forty years, specifically the future of the two-cylinder engine. After a lengthy and spirited debate, now a full decade into the tractor question, the board came to an agreement on the future of the John Deere tractor:

1. The International Harvester Titan and Fordson are "our main competition."
2. A new tractor should be a three-plow tractor meeting the requirements of 75 percent of that trade, based on the same way that the Fordson met 75 percent of the two-bottom plow trade.
3. A new tractor should sell for $1,000 "under normal conditions."

Just when normal conditions would return was anyone's guess at that point. Sales at John Deere came in at 40 percent of normal volume by the end of 1922, although slight improvements did begin to show. At least fifty-five of the ninety all-wheel-drive tractors built by Deere in 1918 still remained in inventory as well. In January 1923, its list price was dropped to $750, half its original price. The Waterloo Boy was now also listed for $750 at retail.[2]

In addition to depressed sales, Deere addressed a growing transition to four-cylinder engines on tractors and whether they could even afford to stay in the tractor business. "After a full and general discussion, it was unanimously agreed that the two-cylinder engine, being simpler, sturdier, and cheaper, is the right proposition for tractor use and the right proposition for this company as manufacturers."

Deere executives decided their course would change only if "field service in the hands of the farmer has actually demonstrated beyond a doubt the superiority of some other type."

General Motors joined the price war in a last-ditch effort to keep its tractor afloat, temporarily "reducing their tractors down to below cost." It only delayed the Samson's demise. By the end of the year, General Motors liquidated the division, converting the Janesville factory to production of the Samson truck line. Losses at the tractor operation in just a few short years exceeded $33 million.[3]

In Moline, the acquisition of the Moline Plow Company by automaker John Willys in 1919 infused new capital into the maker of the Universal Tractor, the country's bestselling motor cultivator. In "a big surprise to the entire implement industry," Willys lured George Peek, the former Deere sales manager and recent War Industries Board executive, to the Moline Plow Company's top post. Five months later, he brought thirty-five-year-old John Deere Plow Works manager Harold Dineen, one of Deere's key negotiators with Henry Ford, to join him.[4]

The Moline Plow Company included all of the independent interests of the Stephens family at the time of the acquisition, including the Moline-Universal Tractor factory in Rock Island, Illinois; the Stephens automobile and body plant in Freeport, Illinois; a reaper and binder plant in Poughkeepsie, New York; a wagon plant in Stoughton, Wisconsin; the Monitor drill plant in Minneapolis; and the Acme steel plant in Chicago.

Peek's Moline Plow Company developed a new two-row cultivator, a "Real Cultivator for the Fordson," their No. 26 plow, No. 9 disc harrow,

Power-Lift Lister, or Power-Lift grain drill, all built for the Fordson, to help save the company.[5]

Restructuring began at the Moline Plow Company as the result of the "sudden and unprecedented falling off in sales" at the end of 1920. Since its acquisition by Willys, debt of more than $10 million "above normal" had accumulated, according to Peek, who began a complete reorganization.[6]

By the summer of 1922, the price of the Moline Universal fell to $650, less than half of its 1920 retail price. As the calendar rolled over to 1923, production at what was once billed the largest tractor factory in the world was shuttered.

Not even Uncle Sam could bail the tractor industry out in 1921. In January, the United States Tractor & Machine Company, recently relocated from Chicago to Menasha, Wisconsin, introduced their Uncle Sam 12-20 tractor, a smaller version of their three-year-old 20-30 model. Within two years the tractor line would be abandoned entirely for automobile manufacturing. Those plans were followed by a swift bankruptcy.[7]

While the farm economy seemed to unravel, Henry Ford remained firmly positioned as its savior. Not only would he build a million tractors every year, and sell them for less than any competitor, he continued to convince farmers that the machines were so productive that they would only need to farm for part of the year.

Meanwhile, as the future state of the tractor industry grew increasingly volatile, engineers at International Harvester and John Deere focused on the development of new models that, in time, would deal a significant blow to Ford's tightening grip on the American farm.

19

"STICK TO IT"

Alexander Legge refused to let Henry Ford price his way into greater market share. But there was a point, he knew, when Harvester's long-term tractor operation would begin to suffer irreparable damage. After all, International Harvester had the most to lose. Despite the euphoria constantly surrounding the Fordson, Harvester tractors had provided farmers with reliable power on the farm now for nearly two decades. More than eighty thousand Titan 10-20 tractors, the company's bestselling tractor by far, had found their way to farms in the last eight years. It was a two-bottom plow, ran on low-cost kerosene, and had proven rugged and reliable. But its design, with an appearance more akin to old threshing machines and steam engines than the sleek, compact design of the Fordson, Wallis Cub, or even the Samson Model M, felt outdated.

Deere's Leon Clausen speculated that the Titan's lower price could not hold, and soon "I.H.C. will discontinue its manufacture." In fact, that work was already well underway.[1]

The new International 10-20 was renamed the McCormick-Deering 10-20 in the fall of 1922, just before production began. The first hundred tractors were expected to be ready by March 1923, supplementing the larger McCormick-Deering 15-30, which had already started to "come through the factory in quantity."[2]

Meanwhile, engineers continued to brief Legge on the Farmall. He "thought that if the machines were so complicated that we feared to turn them over to the farmers to operate, we must be a long way from anything that could be adopted for manufacture." They shipped a tractor to an experimental farm in Hinsdale, a short distance southeast of Chicago, so Legge could see for himself. "The small machine . . . was a big improvement," he told them. This version had differential brakes for short turning, a high-arched rear axle with gear drive, a non-reversible seat, and a front-mounted cultivator. It was a near final version.[3]

Even J.F. Jones, the only Harvester executive in opposition to Farmall development, now approved, though he continued to advocate for smaller production numbers. A heavier version was canceled, "condemned while the

This experimental Farmall was a radical departure from existing tractor engineering. This photo was taken in 1922. Wisconsin Historical Society, WHS-24596.

tractor was still in design," according to its designer, Addis McKinstry. In late 1922, Bert R. Benjamin, a man of modest stature, angular face, and firm eyes, transferred to International Harvester headquarters in Chicago to coordinate all Farmall work, consummating the program to build twenty "light-type Farmalls" for 1923. Internally used by the experimental department since 1919, the word "Farmall" was registered July 17, 1923, and approved by the Harvester Naming Committee on February 5, 1924.

Bert Benjamin was the right man for the job. An 1893 graduate of the Iowa State College mechanical engineering program, he immediately joined the McCormick Harvesting Machine Company as a draftsman and designer. In 1910, he was named supervisor for the McCormick Works experimental department and in 1918 helped introduce the first commercial power-take-off, a device that transferred power from the engine to an implement, a third form of tractor power soon to become an industry standard. For now, he continued to pursue a radical concept he first put into a memorandum in 1916, the design of a tractor and compatible implements that was truly a "one man outfit." In 1920, he had told Legge by letter that the Farmall would eliminate the horse entirely.[4]

With an emergence from the agricultural depression on the horizon, the first 1923 Farmall was shipped to the farm of Sam Insull in nearby Libertyville, Illinois, five miles west of Lake Michigan. Insull was no common farmer. The "utility boss of Chicago," Insull was acting president of both Commonwealth Edison and the People's Gas, Light and Coke Company of Chicago, but had moonlighted for nearly twenty years as a farmer, "working out the farming proposition on commercial lines," he said. Hawthorn Farm employed more than fifty people and was so large that it had its own schoolhouse and post office. For now, after twenty years and continual investment in improved equipment and methods, he was still waiting to generate his first profit. "If farming cannot be made to pay on this place," he told a Chicago paper a few years later, "how can other farmers expect to make money?" He thought the Farmall would help change that.[5]

With a few refinements, the Farmall was ready for its modest debut. Twenty-six tractors were built in 1923, "followed constantly during the season" by Benjamin.[6]

In comparison to its pre-war operations, International Harvester was a very different type of company in 1923. The dismantling of its operations by the Federal Trade Commission had proven to be a pre-depression blessing, forcing the company to streamline its operations and launch a more consolidated brand. Continued diversification and global expansion, with half of its sales now earned outside of the United States, gave aging product lines a longer life in less developed parts of the world.

In July 1923, a petition in the United States District Court at St. Paul, a condition of the original judgment, triggered a review of the 1918 case against Harvester. The petition claimed that Harvester had not moved quickly enough to sell off parts of its harvesting business and that it was now injuring competitors by under-selling them. Its market share had fallen from near 90 percent to a reported 66 percent since the previous judgment.

"Without attempting to argue our case at this time, it seems fair to say that the unfortunate condition in which manufacturers of harvesting machines, and in fact of all other farm implements, find themselves, is due to the serious business depression that affected the entire country beginning in the fall of 1920," Legge offered in a statement. "We are confident that the hearing on the present petition will clearly show that the Harvester Company is not in any way responsible for these conditions in the farm implement industry."[7]

The first half of 1923 showed modest improvement, but lower grain and livestock prices and the closing of a growing number of rural banks encouraged another falling off in the second half of the year.

Internally, despite the introduction of the two-plow McCormick-Deering 10-20, "designed for long life, economy, full power, and operating comfort," Harvester still heard a "constant cry for a tractor to meet Ford's." The 10-20 tractor simply cost too much compared to the Fordson and would never beat it on that point. Something more radical was necessary.[8]

Legge told company engineers that to ramp up production of the new McCormick-Deering 10-20 tractor, the manufacturing department could not afford to give assistance on further Farmall development. J.A. Everson, manager of Harvester's US sales operation, expected the Farmall to fail. Their motor cultivator, designed to plant and cultivate corn, could perform the same operations. It was a better design, he thought, and still it did not sell. The Farmall was too heavy and looked like a dated, three-wheeled tractor. Besides, it would cannibalize sales of the new 10-20 tractor in which they had so much invested.

Legge had actually grown enthusiastic about the Farmall, but he grew increasingly impatient with the business climate, which had yet to offer clear signs of recovery. He also considered the continuous re-designing from his experimental teams an indicator that its designers too were not convinced of its merit. The latest were attempts to reduce the Farmall's weight.

Harvester had been too quick to market with other tractors, Legge wrote, but "this is not true of the Farmall, as you have been building it up and down for five or six years and by this time ought to have pretty fair judgment as to what is the best and cheapest plan to follow."

Plowing with a McCormick-Deering 10-20, 1923.
Wisconsin Historical Society, WHS-24591.

The prospects of a unique system of implements for the Farmall, the creation of a complete farming system for planting and cultivation, was indeed a radical departure from the standard plowing tractor of the Fordson mold. Legge determined it was time to "stick to it" so internally everyone could see one version of the tractor. If it was right, they could move ahead.[9]

By the end of 1923, twenty-six Farmalls were in operation. Half were shipped to farms in Illinois—four to Harvester experimental farms, two to Insull's farm in Libertyville, one to a Mr. Utley's farm, and one to Cyrus McCormick's farm. Thirteen were shipped to branch houses for testing. Years later, only one report was extant, from Hutchinson, Kansas—"and the experiment there had not been favorable."

With continued refinement, two hundred Farmalls were approved for 1924. The first, to avoid mix-up with the dozens of experimental models, was numbered QC-501, and shipped to Taft, Texas. By the end of July, 111 of the 205 eventually built that year were sold.[10]

It was a peculiar-looking tractor, tall, narrow, and almost petite in appearance. Visually, it was certainly more akin to the motor cultivator than its bulkier and more powerful-looking cousins, the McCormick-Deering 10-20 and 15-30. Yet the design suited its purpose. The Farmall was tall to clear emerging corn and cotton crops. Its narrow front wheels, coupled with large, widely spaced rear wheels, allowed for tight turning and operation in rows spaced anywhere between 22 to 42 inches apart, the latter a more standard spacing based on the hindquarters of a mule.[11]

In time, the awkward-looking machine would change farming. For now, it was another experimental machine that E.A. Johnston, Bert Benjamin, and other Harvester engineers hoped found more success than its motor cultivator ancestor.

20

"A VENGEANCE"

Tractor production held less and less of Henry Ford's attention while they rolled out in record numbers from the River Rouge. Nearly 58,000 Fordsons emerged through the first half of 1923, nearly twice the number built in all of 1922. More than 100,000 were built by the end of the year, or a staggering 76 percent of all tractors built in the United States that year. Rock-bottom prices continued to attract first-time buyers. Ford partially attributed the increase in sales to tractor adoption off the farm.

A fleet of tractors was being used for grading work at the new 25,000-seat Bucknell University Stadium in Lewisburg, Pennsylvania. The Street Department in Richmond, Indiana, bought five Fordsons to replace six wagons and a three-ton truck. The city of Lynchburg, Virginia, replaced three horses and three mules, and the tractors were "high in their praise."[1]

C.L. Brown, on behalf of the Mexia Torpedo Co., in Mexia, Texas, bought a Fordson to work the oil fields. He "voluntarily" told the *Ford News* that "with my Fordson I pull casing, move machinery, pull derricks into place, pump mud for plugging purposes, and set off shots of glycerin with which I shoot casing, as successfully as a steam generator or electric battery would do."

While on a branch visit in Los Angeles, Edsel Ford stopped by the Pickford-Fairbanks Studio on Santa Monica Boulevard and Formosa Avenue, the studio of Douglas Fairbanks, fresh from starring roles as Zorro and as D'Artagnan in *The Three Musketeers*, and his wife, the prolific actress Mary Pickford. A Fordson tractor was employed moving sets during the production of a new feature film, *Douglas Fairbanks in Robin Hood*, one of the most expensive films made to date.

In Bicknell, Indiana, a Fordson helped excavate a shaft, then operated a hoist to save a 700-pound pony named Rex. "The age of machinery has arrived with a vengeance," concluded the *Oakland Tribune*'s coverage of a week-long Fordson demonstration in the summer of 1923.[2]

An economic recovery in 1923 finally allowed Ford to take full advantage of the massive complex at the River Rouge. Ford's ambition of one million

The Ford assembly line, like this one at the River Rouge in 1923, transformed the tractor industry. Image from the Collections of The Henry Ford.

tractors a year did not seem so out of reach. Nearly two million Model Ts rolled off the assembly lines at the River Rouge by the end of the year.

In October, surely to Edsel Ford's relief, Henry Ford increased the price of his tractor from $395 to $420, a signal to competitors of better times ahead.

Worldwide, Ford's manufacturing capacity continued to expand. Factories and distribution centers were never finished, being constantly built, closed, and relocated, all in the name of optimization. The once-lauded Fordson factory in Cork, Ireland, at last completed its first tractor in July 1919, assembled 3,626 tractors in 1920, but managed only 1,443 in 1922. Supplemental work building engines and axles for the Manchester factory could not save many of the eighteen hundred employees working in Ireland. Pressure from the Irish government, which included new tariffs, caused the relocation of automobile parts to yet another new facility outside of Cork. Fordson production continued to limp along, just as it took an historic leap in the United States.[3]

The Model T was assembled and sold by twenty-nine branches around the world by the mid-1920s. An assembly facility was enlarged in Copenhagen in 1923 to build six hundred cars a day, and factories in Spain, Italy, and eventually Germany, by the end of the decade, continued to buoy sales. A factory outside of Paris, in Asnières, was operational by 1925. By 1924, Ford branches in São Paulo, Brazil, and Buenos Aires, Argentina, were each assembling and selling 45,000 cars a year.

In the USSR, Ford had taken on folk status that possibly even exceeded his status in the United States. Towns were named after the Fordson tractor, as were children. From 1920 to 1926, nearly twenty-five thousand Fordson tractors made their way to Russian farms, and in turn, dozens of Soviet engineers traveled to Ford factories to study, train, and learn how to develop "Fordizatsia" in the USSR.

The Model T had reached the far corners of the world, and signs indicated that the Fordson was not far behind. The government of Japan bought three tractors, with more orders to come. In the summer of 1923, one hundred students from China, escorted by Dr. Joseph Bailie of Peking University, completed the Ford Service Course at Highland Park, learning about the Model T and the Fordson.[4]

More than 83,000 Fordson tractors were built in 1924, jumping to a record 104,248 for 1925. The industry as a whole also turned in a record year, with an astounding 164,000 tractors built by a shrinking number of manufacturers.[5]

But pockets of growth masked a reality that Henry Ford was unwilling to accept. Sales were stellar, but market share was shrinking, falling to 61 percent in 1925, its lowest since the Fordson was introduced in 1918.

Model T sales too were suffering from new competitive offerings, not only engineering advances, but marketing innovations as well. Sales at Chevrolet had grown from 280,000 to 470,000 automobiles from 1924 to 1925, and would explode to 730,000 by 1926. As a result, Ford's market share in the automobile industry was also falling, from 57 percent in 1923 to a low of 47 percent by 1925.

❧ ❧ ❧

On more than one occasion, in private, Edsel Ford and his brother-in-law, Ernest Kanzler, worked on a redesigned Model T. Henry Ford wanted nothing to do with the notion and rejected even conversation on the topic time and time again.

On one occasion in 1924, Ford walked into a garage at Highland Park upon his early return from a two-week trip abroad. Sitting there was a bright red Model T, its body redesigned by Edsel and a small group of engineers. Ford "walked around the car three or four times, looking at it very closely," wrote George Brown, the purchasing agent who procured the parts. He calmly looked across the hood, down the side panels, and as his anger boiled, he ripped the driver side door from its hinges. "God, how the man done it, I don't know," Brown recounted in disbelief. Ford sat inside for a moment, his mind furiously contemplating the details of the streamlined interior. He ripped out the passenger door, smashed the windshield, and continued to wreck the car until little glass or undamaged steel was left. Henry Ford did not like surprises. He especially did not like anyone, especially his son, who had tried for years to introduce new models into the Ford lineup, to consider changes to the perfect automobile.[6]

"With every additional car our competitors sell," Kanzler persisted still a few years later, "they get stronger and we get weaker." Ford responded with a personal attack, cut Kanzler out of meetings, then fired him.

Pressure mounted as the industry inched forward with them. Ford's ever-reliable strategy, a price cut combined with increased production, was not working.[7]

21

"HIGH HOPES"

Americans continued to be seduced by the consumer-based society Ford helped create with the seemingly simple idea of an affordable car for the masses. But tastes were evolving. Now seventeen years into his tenure as John Deere's president, William Butterworth told his board that farmers were buying more automobiles, radios, electric washing machines, and other luxuries. Their customers' money was being redistributed, and there had been "financial embarrassment of some of the implement concerns" in recent years. Deere "without a question of doubt . . . will come through the period of depression in good shape," he assured them.[1]

Twenty of the leading farm equipment manufacturers lost more than $50 million in 1921 and 1922, Butterworth told listeners of a radio address broadcast in Chicago, Omaha, and Kansas City in early 1924, and the depression came without warning. Yet, "the farmer and the farm equipment industry now face a period of better times," he offered optimistically. He was, of course, always bullish on farm equipment. "They are wealth-producers, cost reducers that saves man—labor costs in the field soon pays for itself."[2]

The implement trade for 1924 was one of "high hopes and very modest realization," according to the *Farm Implement News*. Increases in dollar sales were recorded in most categories—tractors, combine harvesters, corn planters, hay loaders, wagons, and more—but production fell in most areas.[3]

Four out of five of the 230,000 farms in the state of Illinois had an automobile on them. Nearly 12 percent of farms had electricity, and 6 percent had complete plumbing systems. More than 68,000 farms, nearly 30 percent, now included a tractor. A few even had more than one.[4]

A slow revival in Waterloo Boy sales began after its 1921 low point, but the tractor felt outdated and had maximized its power capacity. Now, just as International Harvester began development of its radical new tractor, the Farmall, Deere returned to a prototype developed by Lou Witry in Waterloo prior to Deere's acquisition of the Waterloo Gasoline Engine Company.[5]

While International Harvester focused its new tractor on corn farming, Deere's research pointed to a growing need for a tractor designed specifically for grain-growing sections of the country, which had been shifting from the east to Kansas, North and South Dakota, Nebraska, Oklahoma, Illinois, and Missouri. Deere considered a tractor for these markets a greater opportunity than one for the Corn Belt. Eighty percent of the wheat in the United States was being produced at an elevation between 500 and 1,700 feet above sea level. In Canada, the production of wheat doubled between 1914 and 1924. The United States was close behind. A wheat land tractor offered greater savings over horses than a tractor where corn was the primary crop. In grain regions, "the tractor is used to plow, disk, harrow, seed, harvest, and thresh," and actually required fewer horses for the farm. In corn areas, or on more diversified farms, tractors could be used primarily in preparation of the seed bed, plowing, disking, and seeding, but horses were still essential for other operations during the year.

The United States lagged well behind other countries in wheat yield per acre, averaging 13.9 bushels per acre compared to nearly thirty for Germany, and 32.9 bushels per acre in the United Kingdom. "It is very evident that our wheat farmers must increase the efficiency of their soil," wrote Deere's crop expert, Dr. Warren E. Taylor, "or surrender their supremacy in the production of mankind's most important food."[6]

What became John Deere's model D tractor was first conceived in 1917, a year before its acquisition of the Waterloo Gasoline Engine Company, as an updated Waterloo Boy. At the time, they were calling it the "bathtub" tractor, named after its tub-shaped main case, featuring roller chains that ran in an enclosed oil bath. Early tests exposed the gears to an excessive amount of dirt, which the enclosed gear solved. While experimental Farmall tractors, designed for corn, worked experimental farms in Illinois, Deere tested three separate series of experimental models from 1919 to 1922, primarily near its home bases in Waterloo and Moline, then in Jamestown, North Dakota, west of Fargo and north of Aberdeen.

In Waterloo, twenty-eight-year-old Jack Cade assisted long-time chief designer Louis Witry on design. Cade moved to Waterloo, Iowa, in March 1920, with his wife and three children, two years to the month after Deere's acquisition of the Waterloo Gasoline Engine Company. Cade had been working at the Dayton-Dick Company in Quincy, Illinois, builders of the Leader tractor, their largest being an 18-25 model. Before that he worked for the Hart-Parr Company, the company that cleverly coined the term *tractor*.

An experimental Waterloo Boy tractor, which later became the model D, 1919. Courtesy John Deere.

The third experimental model D was given increased engine speed, from 750 to 800 rpm, and the cylinder block was attached to the front end of the main case. The front axle and radiator were "carried on a frame which extended out from the bathtub case" to avoid a Fordson patent. After investigation, they found that Ford would not contest a replication of the front-end design, and the experimental D was again modified to its original configuration. "It took a lot of the cost out of the job." The tractors, called #301 and #302, were also shortened so they could be more economically shipped crosswise in a boxcar.[7]

On its way to a $430,000 operating loss at the Waterloo factory, Leon Clausen pushed for ten more experimental models in early 1922, suggesting consideration of a four-cylinder engine. C.C. Webber, whose Minneapolis sales branch was responsible still for selling the remainder of Joseph Dain's four-cylinder, all-wheel-drive tractor, advocated for the two-cylinder design well proven by the Waterloo Boy. Now was not the "opportune time to spend any amount of money experimenting with a four-cylinder tractor," he wrote.[8]

Farmer H.G. Glessner ran a model D tractor on his 180-acre farm beginning in July 1922 and gave encouraging feedback. "The tractor turns short, steers easy and my eleven-year-old son does the work whenever he is home from school. He can handle the tractor perfect, and I feel safer to send him into the field with this tractor than with a team of horses."

Glessner found it "economical on kerosene and oil" and without "fan belts or water pump to give trouble."[9]

Howard Railsback, Deere's public relations manager, talked to a former schoolmate, "quite a John Deere booster," about the merits of the new model D. He found his Fordson to be underpowered but was in disbelief that the new Deere was a two-cylinder tractor. "You really mean that John Deere is going to build a two-cylinder tractor?" he questioned.

Only later did Railsback understand the hesitation, as Deere purposefully resisted the trend to "follow the procession of competitors into the four-cylinder field . . ." Deere instead considered the two-cylinder engine to be a differentiating feature, "the one principle against which the entire field of competition has set its solid phalanx."[10]

In early January 1923, at the urging of Leon Clausen, Deere's board approved an additional $50,000 to build 850 model D tractors to supplement the 1,550 Waterloo Boy tractors already scheduled. Deere's year-long build schedule would equate to what Henry Ford built in a just a few weeks. But compared to a paltry seventy-nine tractors sold in 1921, the prospect of selling more than two thousand tractors in a year was encouraging—yet still far from profitable.[11]

Replacement of the Waterloo Boy was officially underway. They projected it would take six months to retool and rebuild the production line and its processes and another two months to build the first fifty tractors. All signs "indicate that the new tractor will be better than anything on the market . . ." Charles Deere Velie pushed for aggressive sales "to secure a proper representation in the tractor and stationary engine field."

George Crampton, Deere's long-time treasurer and even-longer-time advocate of standardization, opposed the new tractors for 1924. The risk was disproportionate to the potential profit, he argued. "If we had made ten Dain tractors instead of one hundred," he reminded them, "we would have saved a large sum." The Janesville Machine Company, owned by General Motors, he added, lost $33 million in the tractor business in just a few years, and "many concerns have gone on the rocks" for taking undue risk in the tractor business.

Clausen disagreed. They were building an entirely new line of business, not just a tractor, he argued. The model D weighed two-thirds as much as the Waterloo Boy N it would replace but achieved greater drawbar horsepower. And it could compete with the comparable McCormick-Deering 15-30, sales of which were growing steadily each year. He had direct knowledge that International Harvester had spent more than $250,000 to rearrange their Chicago Tractor Works for the new McCormick-Deering 10-20, its four-cylinder answer to the Fordson. For the moment, Deere was making a much smaller investment. "I feel that all of the members of our Board of Directors do not realize how good a machine we have," Clausen wrote, inviting them all to visit Waterloo to see for themselves.

Model D production was increased to one thousand units for 1924, the first full year of production. Furthermore, Deere leadership, most at least, reasoned that if they continued to build a line of implements compatible with the Fordson, their own tractor could co-exist, filling additional gaps for customers while treating the Fordson as an extension of their own product portfolio. They could keep pace with manageable plow volumes for Ford as part of its overall business, not as a single source like some of its smaller, less-diverse rivals.

❧ ❧ ❧

Theo Brown returned to Dearborn in April 1923, joined by William Butterworth's nephew, Charlie Wiman. On their arrival, they met Ford, Galamb, Farkas, and others to finalize machine design and terms on the No. 40 Plow. The once-promising partnership had stalled but was revitalized in step with the strengthening economy.

Ford's experimental farm in Dearborn had become a hub for equipment manufacturers, all looking for insights into development opportunities, and ideally a blessing from Henry Ford himself. Former Deere executives George Peek and Harold Dineen, now both of the Moline Plow Company, were there when Brown returned on May 1 to follow up on work that Ford had promised on the Fordson's drawbar, an accommodation for attaching the Deere plow with its self-leveling hitch.

On both occasions, Ford was in a sentimental mood, the meetings feeling more like the end of a relationship than the beginning of one. One hundred No. 40 Plows were in the hands of farmers, and Ford thought it "the best plow he had seen," adding that "he thought perhaps we were making a mistake in not building thousands right away."

Like two scheming schoolboys, Ford and Theo Brown sat together on a rail track "for a long time" and talked. Brown worked up the courage to ask for an autographed copy of Ford's autobiography, *My Life and Work*. Ford asked, in return, for a portrait of John Deere and a "relic" of some kind. "He said it was sentiment, he supposed."

John Deere advertising featured the No. 40 and the No. 41 tractor plows. This brochure from 1923 features a Fordson in front of Henry Ford's boyhood home. Courtesy John Deere.

During his May visit, Brown asked a second time for an autographed copy of Ford's autobiography, which Ford again promised to send, though he no longer needed help with a portrait of John Deere. He found one on his own.[12]

⸙ ⸙ ⸙

International Harvester sold fifteen thousand tractors in 1923, "which was not a good year," Deere's Leon Clausen shared with the board. Harvester appeared vulnerable as its transition to the new McCormick-Deering line had consumed considerable resources. "We certainly ought to be able to sell one-third as many as the I.H.C. If we do not do this, it seems to me we are not really in the tractor business."

In the field, Deere's model D saw immediate results, and for at least the next year, Waterloo Boy and model D tractors were sold side by side. Deere recognized that the Waterloo Boy was "a little out of date in symmetry of design and we know it is a little heavy according to the whims of fashion." According to the sales program, Waterloo Boy N tractors and model E engines would be sold through February 1924, after which the model D would become the featured tractor.[13]

In July, fresh off the launch of the model D, Leon Clausen's new position as president of the J.I. Case Threshing Machine Company was announced.[14]

The move had implications for Deere's tractor business, not only because Clausen's tenacity and advocacy for the tractor business would be missed, but also because of his intimate knowledge of designs, production volumes, and marketing strategy. J.I. Case, the long-standing leader of the steam traction engine industry, had found some success with several tractors built for them by the Minneapolis Steel & Machinery Company, builders of the Big Four 30. From 1912 to 1918, Case offered both a 12-25 and a large 40-80 tractor, the latter rated to pull four plows or more. In 1915, Case introduced its answer to the Bull, the four-cylinder Model 10-20, and gave a strong showing with sales of more than 6,600 tractors over the next three years.

Clausen looked to help J.I. Case find its identity. Two companies with the Case name continued to operate, the J.I. Case Threshing Machine Company and the J.I. Case Plow Works. The J.I. Case Threshing Machine Company sold Case tractors and Grand Detour Plows (a company it had bought, a descendent of John Deere's original plow company), while the other firm sold tractors under the Wallis name and plows with the Case name.

Everyone was confused. The Wisconsin Supreme Court ruled on what products could carry the Case name, and the United States Postmaster General created a process to decide who read what mail. If sent without a street address, which was most mail still, letters addressed to Case had to be opened at the post office in the presence of a representative from each firm. Disputed mail was to be reviewed by the court to determine who could respond.[15]

William Butterworth tapped his thirty-two-year-old nephew, Charles Deere Wiman, to replace Clausen as the company's new director of manufacturing. Wiman quickly made the role his own.[16]

Charlie, as his friends called him, was born on Staten Island. His mother, Anna, the youngest daughter of Charles and Mary Deere, died in her early forties when he was just fourteen. His grandfather died the year after. Charlie and his brother Dwight moved in with their grandmother in Moline. She died in 1913, followed by their father, William, in 1914. The two young men inherited their grandparents' mansion, across the street from their childless uncle William and aunt Katherine Butterworth, but they were rarely there.

The manufacture of farm equipment was Wiman's birthright, but he was first called to the skies. Like his uncle, fellow Yale alum Willard Velie, Wiman was early on seduced by technology. In his case, it was the worldwide obsession with human flight that captured his imagination, fueled by successful demonstrations in France and closer to home by Wilbur Wright's successful flight above New York in 1909. From a family of means, Wiman could act on his dream, beginning instruction in Mineola, New York, under the direction of Howard Rinehart, one of the original members of the Wright Flying School. At one point, prior to American entanglements with Mexico in World War I, Rinehart delivered and demonstrated planes for notorious Mexican bandit Pancho Villa.

Wiman first joined Deere upon his graduation from Yale in 1915, earning 15 cents an hour at the Plow Works. A year later, he enlisted in the United States Army, reporting first to the Governors Island Training Corp. Training began in a Wright Model B biplane, Wiman making his first solo flight in the summer of 1916 before graduating to a Curtis-JN4 under the direction of Filip A. Bjorklund, a Swedish pilot trained in England. The JN4, an unarmed training plane, was capable of speeds up to 80 miles per hour.

After more than one hundred training flights, injuries from a crash in September of 1916 ended Wiman's flying career. His co-pilot was killed, and

Wiman returned to civilian life in Deere's sales department. As soon he was able, he reenlisted in the regular Army just before the United States entered the World War. He served eleven months in France with the Third Artillery of the Sixth Division, discharged with the rank of captain.

At the conclusion of the war, Wiman rejoined Deere for good. He advocated strongly for research and development, especially in the tractor and harvester lines. He also carried the trust of both the men in the shops and the executive team in Moline.[17]

Similar to the reception Butterworth received when he was named Deere's president in 1907, outsiders considered Wiman a strange fit for the agricultural equipment manufacturer, despite his lineage to the company's founder. Unlike his more senior colleagues, he did not grow up on the farm, nor did he actively operate one. *Fortune* magazine found disparity between Wiman and Butterworth, the latter owner and operator of three farms. Burton Peek, Willard Velie, and other executives ran experimental farms as well. Instead, Wiman came to operate two yachts and could even be found enjoying an alcoholic drink

John Deere model D tractor and grain binder, 1924. Courtesy John Deere.

on occasion. Butterworth preached temperance and gave liberally of his time and his money in a very public way, most notably to the Boy Scouts, where he served as a national officer. Moline residents were "puzzled" by Wiman, who kept his philanthropy private.[18]

Beginning in late September 1924, Deere closed its recently enlarged factory site on Miles Street in Waterloo for three months to prepare for increased production of the model D. Three thousand tractors would be available for 1925, a modest goal compared to the McCormick-Deering 15-30. The number, though, just might offer something that had alluded Deere to date in the tractor business—a profit.

22

"POWER FARMER"

Chicago's *Journal of Commerce*, which closely tracked International Harvester's every move, ranked farm implements first in its listing of the three industries suffering the most in the mid-1920s, the others being hide and leather and chemical and fertilizer. The Avery Company, managed by long-time National Tractor Demonstration chairman J.B. Bartholomew, filed for bankruptcy in 1924, though it would be reorganized and survive a bit longer. The Advance-Rumely Company, long-time makers of threshing equipment, steam engines, and in recent years a line of Oil-Pull tractors, reported a $250,000 loss. The cost of development was indiscriminately consuming businesses that had been in operation for generations. Firms were over-extended, each trying to develop into a "miniature International Harvester Company." Many invested everything into the tractor business, only to find that "uncounted acres all over the country were covered with rusting tractors."

The export market for American farm tractors continued to develop. Joseph Stalin's post-war USSR was swept by famine in the early 1920s, but found support in more than $20 million worth of corn and wheat seed sent by the

United States in 1921. Stalin began buying American tractors in 1925, purchasing more than 27,000 machines by 1927. The United States Department of Commerce asserted that the "supremacy of the American tractor throughout the world is undisputed." More than fifty thousand American tractors, nearly a third of all production, were exported in 1926 alone.[1]

Engineers, salesmen, and even executives at International Harvester continued to be encouraged by field tests of the Farmall. For now, growing demand was a problem without a solution. They had no facility in which to build the Farmall in significant numbers.

In the summer of 1925, J.L. Hinton, an engineer at Parlin & Orendorff, the plow company Harvester first purchased in part to keep Deere out of the harvesting business, urged the production of a four-row planter based on recent experiences with users in Texas. "The Farmall is becoming very popular," he wrote the home office, "and they all want to use them for planting as well as preparing the land." Tests were showing a newly designed four-row planter to

Harvesting corn with a Farmall tractor and corn picker, 1924. Wisconsin Historical Society, WHS-23602.

be successful, planting forty to fifty acres daily. "A four-row planter is in very great demand."

The Farmall was still being sold on a demonstration basis only. A tractor would be taken to a farm for a week or so, the owner told he could use the tractor for any operations he wanted. "Very rarely was a Company man permitted to take the tractor off the prospect's farm after he had used it," it was later remembered.

Farmall number QC693 completed thirty-nine hours of testing by the Nebraska Tractor Test Laboratory over a five-day period, from September 14 to 19, 1925. "There were no repairs nor adjustments necessary during this test. At the end of the test the tractor was in good running order and there were no indications of undue wear nor of any weakness which might require early repair."[2]

The Omaha branch was still instructed not to push their sale. If the tractor sold as well as they expected, Harvester did not have the facilities to build more.

But they were working on a solution. With the Universal tractor out of production, the assets of the nearly defunct Moline Plow Company were sold to the Moline Implement Company, a holding company. It included the 425,000-square-foot tractor-assembly plant, which was put on the market and quickly liquidated. The buyer was International Harvester.

The *Moline Daily Dispatch*, Deere's hometown newspaper, broke the news: "Harvester Company Takes Over Giant Tractor Plant."

The factory would be used for storage, the company claimed, soon after noting that it would actually be retrofitted for use as a transfer house, a staging center for machines with destinations across the country. Regardless of its use, another Moline newspaper concluded, the acquisition was yet another indicator that the "days of depression appear to be left behind."[3]

J.F. Jones, the lone holdout in early meetings regarding the Farmall, was now fully behind the new tractor and recommended production of no fewer than one thousand by the end of October and an equal number of middle-buster plows as well. Requests from overseas were beginning to come in, but for now, the Farmall would be available only in the United States. Production targets for 1926 were increased from 1,500 to 2,500 tractors. They expected it to sell alongside existing models without additional advertising.[4]

"In 1922 the Harvester Company brought out the two well-known tractors," an early 1926 advertisement for its 10-20 and 15-30 tractors reminded customers, to "fit the requirements of general farming." They had become the "quality standard of the world, the pride of every owner."

To supplement the line and make "true horseless farming possible," the "McCormick-Deering FARMALL, specially built for planting and cultivating corn, cotton, and other row crops, and at the same time as perfectly adapted for plowing, drawbar, belt and take-off work," was now available.[5]

The Farmall began to appear with the McCormick-Deering models at dealer training schools like the one hosted by Charles Woolworth in Mt. Pleasant, Michigan. A knocked-down McCormick-Deering 10-20 was fully assembled in ninety minutes' time before a crowd of two hundred people "to show the precision with which the parts went together smoothly and accurately." A two-reel motion picture, "A Trip Through a Tractor Factory," was one of several shown in additional lectures.[6]

Harvester was on pace to sell nearly twenty thousand McCormick-Deering 10-20 and nearly thirteen thousand 15-30 tractors, almost double its production from the previous year. Significant advertising resources continued to support the tractors, but there were still gaps in the South, especially in cotton country, where neither model was selling well. The possibilities of the Farmall in these areas, largely covered by the Omaha branch, convinced them of at least limited opportunities for the tractor and its custom line of implements.

John Deere was "better situated than the average of the industry because it went back to a plow business that made the name of Deere famous, trimmed its sails and prepared for a storm," the *Journal of Commerce* concluded in its article on the hard times suffered by evolving industries. The paper was referring to the success of the No. 40 Plow and other implements made for the Fordson, and its overall success in earning business from operators of all tractor brands.[7]

Deere's transition from the Waterloo Boy to the model D had proven successful, and its tractor operations, still modest in comparison to both Ford and International Harvester, enjoyed its most successful period to date. From November 1925 to September 1926, Deere shipped 7,396 tractors. Another 1,200 tractors were carried over to 1927, which "is not considered a very excessive inventory," Charlie Wiman reported.[8]

The model D tractor was, as advertising stated, simple, durable, and as time would show, built for years of work. But the model D was not a general-purpose tractor, and it was clear that Deere still needed a tractor for the Corn Belt to compete with the growing reputation, and sales, of the Farmall. Wiman, who was far from Moline when Theo Brown developed his earliest motor cultivator concepts, asked him to revisit his Tractivator as inspiration for an updated "all-crop" design.

Brown poured himself into the project, completing three working prototypes in the summer of 1926 and delivering field models for testing the following summer.

Internally, John Deere's "all crop" was controversial from the start. The Fordson was a two-plow tractor, as was the Farmall, but Brown pressed for a three-row tractor to separate it from the rest of the field. Two were being tested in southern Texas, in Mercedes, along with an experimental three-row cultivator and corn planter. Its performance "has demonstrated the design of the outfit is satisfactory" a report to the board outlined, but it was too early to make final design decisions. Another twenty-five would be built to test the three-row concept.[9]

Wiman considered the model D "foremost in its particular class," but "in the natural course of events, competitors have made improvements in their tractors." He asked for $50,000 to strengthen gears and shafts from the crankshaft back to the rear axle and hubs, walking the board of directors through the sample floor and pointing out proposed changes, pushing for them to "consider and pass upon the appropriation in an intelligent way." Increased horsepower on the model D would make it even more competitive against the McCormick-Deering 15-30.[10]

With an updated model D tractor underway, focus moved to the race between the Fordson and the Farmall. Internally, Deere now took to calling their "all-crop" tractor the model C. Others called for a more descriptive trade name, some adopting Power-Farmer, at least for now. Ever the optimist, Frank Silloway thought they could eventually realize annual volume of fifteen thousand row-crop tractors.[11]

The Power-Farmer would be a low-cost tractor suitable for plowing and row-crop cultivation, putting it firmly in the general-purpose class, but not necessarily a direct competitor to the Farmall. But the unique three-row concept was a radical departure that would require a separate line of redesigned implements. If testing was a success, Wiman, who had come around to the concept, argued, "it will avoid the manufacture of two models," one for plowing and one for cultivation. The goal was to build a tractor to operate in corn rows grown as high as twenty-five inches.

"Let us for a moment take the worst possible view of the situation, namely, that the four-wheel, three-row method of cultivation does not work out. It will then be necessary to work out immediately a tri-cycle machine like the Farm All, design on which has already been started."[12]

* * *

In 1927, seemingly a lifetime since the price wars began, International Harvester vice president Addis McKinstry, one of the principal Farmall designers, told a group of Iowa dealers that the "country appears to be going ahead soberly, prudently, energetically, and confidently with the wiping out the last traces of the disaster of 1921, and with the still more important job of building for a new and solid prosperity."[13]

The servicing of power farming equipment brought new challenges. "Power units, no matter how well we build them, will always require a degree of follow-up service not called for by the older and simpler horse tools," Legge told a convention of farm equipment dealers, the most receptive audience for such a statement.[14]

Harvester's lineup for 1927 included the McCormick-Deering 10-20, 15-30, and the Farmall. Specialty models were boosting sales as well: An orchard version of the 10-20 and 15-30 featured rear-wheel aprons and a lowered seat and steering wheel to protect branches in orange and citrus groves.

Spade lugs reduced soil compaction, an especially important factor in irrigated orchards. Narrow-tread tractors could be purchased for row-crop applications, and industrial versions with hard rubber tires, capable of forward speeds of ten miles per hour, were in demand by county and state municipalities.[15]

To help facilitate sales, a new Red Baby, a "flaming red," three-passenger canvassing vehicle with "room for sample engines, cream separators, grinders, twine, and what-not" was available to dealers. "The man in the city seldom stops to note the ramifications of delivery services," an editor in Pennsylvania pointed out in his coverage of Edward Cornish, manager of the Fayette Farmers Supply company. "The farmer is a businessman," and time away from planting or harvesting reduced profit. "Think then what a boon it would be to the farmer that he could telephone his wants in town and have them satisfied by quick deliveries."

More than six thousand Red Babies were already traveling rural roads across the United States.[16]

Harvester branches estimated they could sell ten thousand Farmalls for the 1927 season, more than double the previous year and nearly a third of the number of 10-20 model tractors they soon would sell that year. Legge thought they would have to "get up our courage and perhaps go to a bigger schedule in order that an accumulation of Farmall tractors could be made . . ."[17]

A work force of more than six hundred began Farmall production in Rock Island at the newly renamed Farmall Works, on June 21, 1926, growing to nine hundred the following spring after Harvester announced the purchase of additional property formerly owned by the Moline Plow Company. New facilities would store harvesters and binders less than a mile from the John Deere Harvester Works in East Moline.[18]

"The tri-cities are in the center of the corn belt, and nothing seems more fitting than to have the International locate its Farmall factory here," a spokesman said. "The product may be easily and readily distributed in all directions directly to the corn growers." The plant, the paper called out in the headline, was one of the world's largest.[19]

In the spring of 1927, from California, president Alexander Legge urged an increase of Farmall production to fifty tractors a day. It would take five weeks from the order, he learned. Factory enlargement could not keep pace with orders. What about forty, he responded, via telegram. On one day in October, they built ten. Two weeks later, twenty. On November 14, 1927, the crew set a record, building sixty Farmall tractors in a single day.[20]

"Necessity and the changing times have brought about great improvement on the farms," stated a Harvester newspaper advertisement. "The young farmer and the old farmer who keeps his mind young are working on a new and profitable scale. The McCormick Deering 10-20, 15-30, and the new FARMALL—is evidence of the new era of farming."[21]

23

"THE LAYOFF WILL BE BRIEF"

Henry Ford continued to sell cars by the millions and tractors by the tens of thousands. But both markets had evolved without him, and market share continued a downward trend. In 1927, Chevrolet overtook Ford on annual sales of one million cars.[1]

Edsel Ford, now only thirty-four years old but burdened with years of discouragement and opposition, finally convinced his father that the Model T's run was over. Speculation about a new car circulated for months, and the Ford team deployed one of their best-tested tactics, leaking intermittent enticements to excite the public on the offering soon to come. "The 'mystery' over the Ford model was calculated to induce prospective customers to wait before buying other makes, but it was decided not to rely on this entirely and quantity production was rushed slightly faster than was planned," one newspaper offered in May.[2]

In reality, reminiscent of his hesitancy to release his tractor for production and shipment to the British government, Ford could not let the Model T go, refusing time and again to approve production of its replacement until it was perfected. And only he could judge when that time had come. Even the agency hired to advertise the now much-anticipated automobile was dumbfounded. After a visit to Detroit, a copywriter found the car "in the middle of the floor, nearly finished so far as we can make out." Each day the engineers would gather around it, waiting for Ford to appear and give his approval, but each time he found some minor, often imperceptible fault. "With that he walks off, leaving them there gazing at the car. It's like that day after day, with the whole country screaming for the new Ford car. We don't understand it," he told his bosses.[3]

Edsel was eventually able to push things forward. Three days after Detroit native Charles Lindbergh successfully navigated his single-engine monoplane, the *Spirit of St. Louis*, on the world's first non-stop transatlantic flight, Ford sought to recapture the headlines with announcement of the coming of the first all-new Ford automobile since 1908. The Model A was not yet ready for production, but dealers began clearing Model T inventory in preparation. In a well-orchestrated ceremony held the next day, father and son drove the fifteen millionth Model T off the Ford assembly line at Highland Park.

Henry Ford put on his usual good show for the crowd. "We began work on this new model several years ago," he shared, but "the sale of the Model T continued at such a pace that there never seemed to be an opportunity to get the new car started." They had even figured out a way to briefly lay off only twenty-five thousand employees, instead of seventy-five thousand as they had estimated for the changeover to the new model. "The layoff will be brief, because we need the men and we have no time to waste," he shared triumphantly.

The new car required retooling and a complete shut-down at the River Rouge and forty-seven other assembly plants in the United States, Canada, and overseas.

Ford still knew how to give the people what they wanted. "At present, I can only say this about the new model. It has speed, style, flexibility, and control in traffic. There is nothing quite like it in quality and price. The new car will cost more to manufacture, but it will be more economical to operate."[4]

* * *

J.G. Starr and Sons, Ford dealers in Decatur, Illinois, patiently waited through months of misdirection and rumor for the arrival of the new Ford. They planned an exhibition at the Macon County fair, now a week away, but had no word from the Ford Motor Company. Production was to begin in June, then July, now reports said September 1, which would still allow time for the dealership to get one for the fair. "In the event that the new Ford is not available the display of the local firm will consist of Lincolns and Fordson tractors."[5]

In supposedly leaked photos, it was unclear if the new car was a four-, six-, or eight-cylinder model. Then in new photos, the "car was in exact reproduction to scale of the Lincoln. The next it was quite radically different."

"Finally, the public became groggy," offered the *Sunday Star* in Washington, DC, in November. Ford, the reporter discovered on a fact-finding visit to Detroit, would not approve production until they had enough capacity to build three million cars annually.[6]

The much-anticipated Ford Model A finally debuted across America on December 2, 1927, including in Decatur, Illinois. The press, prone to exaggeration when it came to crowd estimates, gushed that ten million people visited showrooms across the country within thirty-six hours of its arrival. Fifty-thousand cars were ordered in New York on the first day, with nearly half a million on order by the end of the year. The lone car that came to J.G. Starr & Sons, which arrived well after the closing of the county fair, was fawned over by seventeen thousand people on its first day of exhibition. That was nearly 25 percent of the city's population.[7]

❧ ❧ ❧

Edsel, who even his father recognized had "mechanical horse sense," easily saw the similarities between the Model T and the Fordson. Strong sales continued, but so did erosion of its market share, steadily falling almost 30 percent over the previous four years.

Small indicators of eroding share mounted. An order of 360 Fordsons from the Department of Agriculture was completed late in 1927, part of a program to destroy the scourge of the corn borer, a pest that had infiltrated

American crops, and was "a tribute to the sturdy and dependable Fordson, which has done so much to remove drudgery from the farm and CUT PRODUCTION COSTS."

In fact, the Fordson order was the smallest of the lot. Four-hundred forty-four McCormick-Deering 15-30s and the same number of John Deere model D tractors, specially outfitted with a special power-take-off for the operation of a stalk pulverizer, rounded out the order. Tractor and implements always went together.[8]

For now, the Fordson kept parts of the River Rouge humming as tractors continued to roll off the line through the summer and fall of 1927. There would be plenty for the spring trade, and farmers could now take "two falls to pay" under a new payment program. With only a small amount of money down, half of the remaining cost could be paid in the fall of 1928 and the balance in the fall of 1929.

"Farming after all is manufacturing," a Fordson advertisement in the fall of 1927 offered, "and just as the manufacturer has adopted the most modern equipment available to speed up and lower his production costs, so of necessity the farmer should adopt Fordson equipment which is generally recognized as another step forward in progressive and profitable farming. The sooner, the better."[9]

John Deere model D and McCormick-Deering 15-30 corn borer tractors await shipment by the United States Department of Agriculture, 1927. Courtesy John Deere.

24

"THE BUSINESS OF RAISING FOOD"

At the outset of 1928, Harvester vice president Addis McKinstry thought it seemed "safe and reasonable to say that the prospects for the farm implement industry are, on the whole, somewhat brighter than a year ago." Increasing sales and market share at John Deere confirmed the trend. They expected to build fifteen thousand model D tractors at the renamed John Deere Tractor Company in Waterloo in 1928, easily its most ambitious schedule to date.[1]

About one hundred tractors of the still-experimental all-crop design, continuing to be called the model C or the Power-Farmer, were in the field, but were plagued by a lack of power and had proven too light in the front end. Engineers were convinced that the problems could be overcome, and "every effort of the Waterloo organization is being devoted to this object." Nevertheless, production would likely be delayed.[2]

Testing continued over a five-day period in early January at the John Deere experimental farm in La Jolla, California. Engineers ran it beside a Farmall and a McCormick-Deering 10-20. No Fordson was there for comparative testing.

The Deere tractor had been elongated to increase its front weight, and the center of gravity lowered by dropping the front and rear ends. The cylinders were increased in diameter for greater power, and the tractor was widened over the front and side hoods to give it "improved general lines." The C performed "almost as well" as the Farmall turning at the end of rows, they concluded, though the Farmall was "slightly more positive" in the control of cultivator shovels in the ground. Comparable work was performed in hillside plowing, with the C testing better in extreme conditions such as the plowing of steep ditches. Allen Head barely escaped a rollover operating the Farmall with its plows out of the ground, while the C proved successful in the same test due to its clutch release feature, allowing him to disengage the plow from a standing position.[3]

Deere began to see its experimental machine as a real competitor to the Farmall. Leadership discussed "where the tractor business of Deere & Company was eventually going," coming to agreement on a "very rapid expansion" and putting plans in place to build twenty model C tractors on top of fifty model D tractors a day by August. That number soon increased to meet sales branch projections for twenty-five thousand model D tractors alone. The Waterloo factory was now in a constant state of expansion.[4]

On January 20, 1928, Deere notified its sales branches that the Ford Motor Company would shut down Fordson production for up to a year to retool. Even on the heels of a similar shutdown to replace the Model T, a yearlong shutdown was suspicious. Production of the Fordson, still mostly unchanged from its original design, reached 93,972 tractors in 1927. Its scale remained the envy of the entire industry. Ford's assembly line put out forty thousand more tractors than the International Harvester lineup combined and double the combined efforts of all remaining manufacturers, including Deere, though to the observant, the staggering figure also signaled a milestone. Fordson market share had fallen below 50 percent for the first time since its introduction in 1918.

"We are told that during this interval, it is their plan to design and prepare for production a new Fordson tractor." In the meantime, Deere instructed branches to connect with their dealers to shore up their inventories and aggressively sell as many existing plows as they could, assuming future incompatibility with a new Fordson. There was now little communication between the two companies. "If Ford should design any new tractor, we do not know whether the No. 40 plow will be the one particularly adapted for it or not. We should operate on the basis that it will not."[5]

The breakneck pace of tractor development was only accelerating. If the Ford rumors were true, demand would far exceed capacity. By late spring, Deere increased production and put in place its largest Waterloo expansion to date. "If the C Tractor rights in its design we will want fifty per day," Silloway urged. "I'm positive that we will require seventy-five Ds per day to take care of volume," sensing, like Harvester's Legge, a need for more, encouraging a "gamble" of one hundred a day.

Deere increased the bore on the model D and replaced the carburetor, increasing power by 25 percent, the equivalent of five to six horsepower. "When the John Deere Tractor was first put on the market, it immediately met with the approval of farmers," advertisements offered. "Its great power combined with light weight; its extreme simplicity; its ease of handling; its low operating expense; its low maintenance costs—all those advantages made it more than a successful farm tractor—it was a real sensation."[6]

Amid ongoing speculation about Ford's future tractor, and far from the fanfare of the fifteen millionth model T the previous year, the last Fordson tractor quietly came off the assembly line at the River Rouge on June 4, 1928. Only eight thousand tractors had been built year-to-date, as equipment was disassembled and crated. Ford advertising from the past year now began to read like a memorial, another effort to forever cement Henry Ford's status as an American folk hero. Ford was the "Abraham Lincoln of the 20th century," a national advertising campaign had touted in 1927, dedicated to the man whose cars and tractors "involves emancipation in another realm."[7]

Lord Percival Perry, Ford's British representative so instrumental in bringing the first Ford tractors to Great Britain during the war, requested and was given permission to ship the factory's equipment to Ireland. "Let Perry have the equipment right away," Ford, who held the sales teams entirely accountable for declining sales, told Charles Sorensen. "They can't have it standing still here. A little later we will do something else."[8]

Temperatures in Washington, DC, reached the upper eighties in the first week of July 1928. Clouds were expected to bring rain the day after sixty-three-year-old William Butterworth addressed guests of the annual meeting of the American Society of Agricultural Engineers. Butterworth, whose hair and mustache had grown silver in recent years, had just been appointed president of the United States Chamber of Commerce. Charles Deere Wiman, the great-grandson of John Deere, would become John Deere's fourth president later that year, though Butterworth would transition to the newly created position of chairman of the board, the "ranking officer of the Company."[9]

The summer meeting was one of reflection and optimism, an acknowledgment that a significant era had passed. The society's president, O.B. Zimmerman, coming from a meeting with presidential candidate Herbert Hoover, reflected on the transition "from old Dobbin to the dependable and versatile tractor." The results were faster farm work, reduction of working hours, and increased production. He encouraged more research, without which "agriculture today would be in much the same condition as it was a century ago."

Butterworth, once considered unfit for his role in John Deere's top position, reflected on the previous eight years. The agricultural population had declined by three million people; the number of horses had decreased by even more, yet the aggregate crop production had increased almost 5 percent. There were an estimated 768,825 tractors now on American farms, an unthinkable number even ten years before. He resisted calling out any specific manufacturers or acknowledging that as many as two-thirds of those tractors were Fordsons.

The Chamber of Commerce had been studying the state of the agriculture industry and concluded that there were "few if any national agricultural problems." But there were eighteen different regional problems identified. Butterworth surely was reminded of Deere's early negotiations with Henry Ford, concluding that his tractor endeavor would ultimately fail because a single tractor could not cater to the diversity of farms across the country.

"I look upon agricultural engineering as the most vital factor in our vitally important agricultural industry," he told a captive and agreeable audience. "Without the research and achievements which have been developed in this field, our agriculture today would be in much the same condition as it was a century ago."

Bumper corn and wheat crops were being reported in Iowa, Illinois, Nebraska, Kansas, Minnesota, and elsewhere in August. Oats and wheat would reach record, or near record, levels as well. The crops in Iowa and Illinois accounted for 25 percent of the nation's corn, with more than 117,000 tractors operating in those two states alone.[10]

Industry insiders were unconvinced that Fordson production had in fact ended, but by summer it was becoming increasingly clear that the rumors might be true.

"We are going to get an increasingly larger percentage of the total tractor business and we should be prepared to take better care of our dealers," George Mixter, who had remained on Deere's board since his departure in 1917, argued, especially with "Ford out of the tractor business . . ." A $4 million construction project began in Waterloo.

International Harvester acquired a block of land between 43rd and 44th streets in Rock Island, Illinois, for further expansion at their Farmall factory, making the site once again, as was for a time claimed by the makers of the Universal, the largest tractor factory in the world. By September, a workforce of eighteen hundred was loading twenty-two train cars with 125 tractors a day. The factory poured 160,000 pounds of steel daily, and consumed enough gas and electricity to power a town of thirty thousand people.[11]

John Deere learned that dealers objected to the designation of the model C, because it and the existing model D "both have the sound of 'Double E.' When a dealer orders a tractor over the telephone, it would be very easy for a misunderstanding to arise . . ."

A new name was adopted. GP followed the alphabetical convention started with the model D, and GP stood for General Purpose, "and our new tractor is a general purpose tractor," it was reasoned.

"Let us never forget that, first, last, and all the time, our tractor is the John Deere Tractor, and the John Deere Company, and all it stands for, is evident

Despite being pitted against each other, total tractor horsepower did not exceed power from horses on American farms until 1945. This John Deere GP tractor and corn picker dumps corn into a horse-drawn wagon. Courtesy John Deere.

throughout its construction." This is why clever names, horsepower ratings, or other gimmicks were unnecessary, they shared with each sales branch.[12]

"As a strong, two-plow tractor, the GP is interesting," offered an industry report in comparison to the Fordson. But lining up against the Farmall, the GP was even more interesting as a cultivating and planting tractor.[13]

"The usual difficulties have been encountered in placing into production this new tractor," Charles Deere Wiman reported late that summer, "but most of these handicaps are being rapidly removed, and it is expected to get into production at a very early date."

Ominously, in mid-August 1928, the city of Fordson, comprised mostly of employees of the Fordson tractor factory, was absorbed by the adjacent town of Dearborn. "The name Dearborn was retained to keep a historically significant name in the history of the middle west," the *Ludington Daily News* explained.[14]

In its first issue of the new year, *Farm Implement News*, which had first shared Henry Ford's ambitions to build a farm tractor twenty years and two months earlier, wrote: "The demand for the standard-type two-plow machines has been filled with no Fordsons in the competitive picture except for second-hand machines and for some few that may have been carried over from the spring of 1928."

"Ford had convinced thousands of horse-using farmers that in the tractor they could find the answer to their power needs," Cyrus McCormick III, who personally oversaw the one hundred thousandth Farmall assembled at the Rock Island factory in 1930, later admitted in a history of his family's century in the farm equipment business. "When the tractor war was over, the farmers of the world appreciated beyond a shadow of a doubt that they would best serve themselves by providing their farms with a tractor rugged enough to resist the shocks of farm use and powerful enough to do all of their work. They knew that there can be no such thing as a good cheap tractor."

❧ ❧ ❧

The American media missed, or perhaps simply ignored, an article published in the *London Times* in late August 1928. "From the standpoint of tractor development, one of the interesting developments of the year was the competition which the American industry was beginning to receive from a tractor builder abroad," Harvester's advertising manager observed. "This development was the inauguration of tractor manufacturing in Ireland, an enterprise Henry Ford developed after he withdrew from the production of tractors in America early in 1928."

Soon after the Ford Motor Company's new automobile, the Model A, reclaimed its rightful place as the world's bestselling car, Henry Ford, a farm boy from Dearborn, Michigan, shared news that automobile production at the factory in Cork, Ireland, the ancestral home of the Ford family, would be relocated to Manchester, England. Instead, the Cork factory would hire forty-five thousand people, who, working three shifts a day, would assemble nearly fifty thousand tractors, mostly for export, the following year. A redesigned Fordson now achieved thirty horsepower and featured updated cooling, lubrication, ignition system, and heavier front wheels for increased traction.[15]

"Power-farming is simply taking the burden from flesh and blood and putting it on steel," Ford, with the clarity of a lifelong mechanic, once said. "The motor car wrought a revolution in modern farm life, not because it was a vehicle, but because it had power. Farming ought to be something more than a rural occupation. It ought to be the business of raising food."[16]

After two decades of farm tractor development, American farmers no longer had to wait for Henry Ford to deliver on his latest promise. The tractor wars were over, for now.

EPILOGUE

Fortune magazine called the 1920s the "decade of the Fordson." Not everyone agreed, but the fact that the Fordson accelerated adoption of the farm tractor and forever altered the competitive landscape is undeniable. Ford was right about the insatiable appetite for automobiles, and in time, for farm tractors. There were now seventeen million passenger cars and nearly 2,500,000 trucks traveling on more than 521,000 miles of surfaced roads. But he had miscalculated the connection between farmers and their land, and their constantly evolving needs. Most importantly, farmers wanted to fortify that relationship, not end it.[1]

Only thirty-three American farm tractor manufacturers remained in 1929, as mergers, consolidations, and bankruptcies continued to narrow the field. In its overview of the tractor industry in early 1928, the *Tractor Field Book* surmised that "large numbers of inefficient machines were discarded during the five-year period, 1920 to 1924" as many were built by "companies whose efforts were largely experimental."[2]

Farm Implement News, which first shared Henry Ford's ambitions to build a farm tractor in 1908, analyzed the impact of power farming on the nation. "Modern power farming with tractors and tools, is a development of the present generation," it editorialized. It now required only ten minutes to produce a bushel of wheat, compared to more than three hours "under old hand methods." Power farming meant greater prosperity for the more than six million farms left shorthanded by the more than four million people who had left the farm for the city in the previous decade.[3]

After decades of development, partnerships, and patience, John Deere was now building nearly a quarter of the nation's tractors. The model D remained in the lineup until 1953.[4]

International Harvester reclaimed its top position, manufacturing 60 percent of the tractors built in the United States. In the spring of 1929, incoming president Herbert Hoover nominated Alexander Legge to chair the new Federal Farm Board, a group convened to administer the Agricultural Marketing Act of 1929, a government effort to modernize the economics of agriculture. Legge accepted the appointment as his civic duty but considered it career suicide.

On April 12, 1930, the one hundred thousandth Farmall, now Harvester's top selling tractor, was assembled at the former Moline Universal factory in Rock Island, Illinois. Two months later, the two hundred thousandth McCormick-Deering 10-20 rolled off the assembly line at the Tractor Works in Chicago.[5]

The Fordson factory in Cork, Ireland, began to export tractors to the United States in 1929, though for a variety of reasons, never reached the capacity Ford promised.

AUTHOR'S NOTE

Fortunately, vintage tractors are studied, restored, and exhibited at thousands of vintage farm shows, state and county fairs, and small-town parades every year. Yet surprisingly, there is often little known about the origins of these machines, why they came into being, or the people behind them. The farm tractor is typically portrayed as an inevitability, like the automobile or going to the moon. Of course, nothing is inevitable until it actually happens.

I was drawn to the story because of the personalities and the brands, and because I wanted to understand why decisions were made and, in some cases, investigated and abandoned. Why did Henry Ford refuse to build implements? Why was John Deere not in the first wave of tractor manufacturers? What drove the design of the Farmall? How were these companies connected, how did they collaborate, and when did they compete? And what were the contextual drivers, because after all, businesses can only operate in a larger political, social, and economic climate.

Ultimately, I hope to shed light on a transformative period in an often-overlooked industry and introduce the people, places, and brands that in many ways defined the century. Today, most of the world does not know where their food comes from, in part because a small percentage of the population feeds the entire world. Power farming made it possible.

Finally, it's important to note research processes and the sources that defined this story. Production numbers and market share are difficult to ascertain during the early twentieth century. Product and model years do not align between competitors, and government statistics rarely match company figures or those reported in trade journals. In addition, it's impossible to normalize reports of tractors built, sold, inventoried, and shipped in model years and calendar years. Although I often cite exact numbers, they are imperfect and are intended to guide the story, not be a definitive analysis on production.

Contemporary sources, including government studies, are the foundation of the book, notably the Theophilus Brown Diaries at the George C. Gordon Library, Worcester Polytechnic Institute, Brown's *Deere & Company's Early Tractor Development*, and *Tractor History*, a manuscript by International Harvester marketing director A.C. Seyfarth. Henry Ford's autobiography, *My Life and Work*, provided insights into his philosophies on rural life and the development of the tractor, and Charles Sorensen's book, *My Forty Years with Ford*, provided a useful chronology and context around the introduction of the Ford tractor in Great Britain, although timelines from different sources rarely agree. Oral history collections at The Benson Research Center at The Henry Ford Museum of Innovation provided critical insights into the development of the Fordson, especially in its first few years. R.B. Gray's two-part study, *Development of the Agricultural Tractor in the United States*, provided insights into the tractor industry on a year-by-year basis.

ACKNOWLEDGMENTS

I am indebted to John Deere, where I have spent much of my career working in one way or another with the corporate archives. I was permitted collection use and, most importantly, encouragement for the work. My agent John Willig, whom I first worked with almost twenty years ago, offered instant enthusiasm for a book about the origins of the tractor industry. Matt Holt, editor-in-chief at Matt Holt Books, recognized the importance of a unique historical business narrative largely set in the Midwest but with global implications, re-energizing my efforts to put this story together. The entire team at BenBella Books, including Katie Dickman, Sarah Avinger, and Brigid Pearson, were responsive and insightful, helping to evolve an exhaustive cast of people, events, and machines into a cohesive narrative.

Foundational information on companies and products came from annual reports, newspaper articles, bulletins, memoranda, correspondence, and diaries. I am impartial, but archivists are the unsung heroes of history, acquiring, managing, describing, and providing access to the primary sources that make a book like this possible. Many of the records I used were secured from online catalogs, and without the ability to visit repositories during the COVID pandemic, several archivists provided generous remote services that went above and beyond.

Arthur Carlson at the George C. Gordon Library, Worcester Polytechnic Institute, supplemented online portions of the Theo Brown diaries, and Sam Julian, Sally Jacobs, Lee Grady, and Lisa Marine assisted with the

McCormick-International Harvester Collection at the Wisconsin Historical Society. Guy Fay, a prolific author of all things International Harvester, promptly offered expertise that would have taken weeks of research to accomplish. Clinton Lawson at the State Historical Society of Missouri helped navigate the George Peek Collection, which includes a trove of correspondence related to the War Industries Board and relationships between farm equipment manufacturers and Henry Ford. Gretchen Small, program director at the Butterworth Center and Deere-Wiman House, shared her unrivaled knowledge of the Deere, Butterworth, and Wiman families. Stephanie Lucas and Jim Orr provided information and expertise from The Henry Ford.

Contemporary newspapers and agricultural and automotive journals provide observations and insights into the world as it was, versus how we want to remember it being. State archives and historical societies continue to add keyword-searchable newspapers from their states, which I supplemented with several subscription services. Digital archives are truly transforming research, and I'm indebted to the work of archives and museums that are digitizing collections and providing accessibility online. This is laborious and, unfortunately, often undervalued work.

Several colleagues allowed me to share early and sometimes unreadable versions of the book, and offered much-needed encouragement. Peter Liebhold, curator emeritus, Division of Work and Labor at the Smithsonian National Museum of American History, is a champion for the role of technology in agricultural history. Dr. Debra Reid, curator of agriculture and the environment at The Henry Ford, is a selfless advocate of agricultural history and technology. Brent Scherr asked essential questions about terminology and equipment, which significantly altered my approach to the narrative. Brian Nicol provided objective organizational advice on methods of handling a sometimes-confusing network of characters and brands.

I must also recognize the farm magazines and the thousands of vintage farm tractor and equipment collectors who keep this era of agricultural history alive. They share their passion for these resurrected iron horses in parades, county and state fairs, school events, and working farm shows. On their own time and at their own expense, they haul their mechanical time machines across the country, sharing insights and memories about this defining era of American history.

SELECTED BIBLIOGRAPHY

Newspapers

Abilene Daily Reporter (Abilene, Texas)
Abilene Weekly Reflector (Abilene, Kansas)
The Alma Record (Alma, Michigan)
Blade (Toledo, Ohio)
Boston Sunday Post (Boston, Massachusetts)
The Clovis News (Clovis, New Mexico)
Cook County News Herald (Grand Marais, Minnesota)
Davenport Daily Democrat (Davenport, Iowa)
Dearborn Independent (Dearborn, Michigan)
Daily Capital (Topeka, Kansas)
The Daily Gate City (Keokuk, Iowa)
The Daily Huronite (Huron, South Dakota)
Detroit Free Press (Detroit, Michigan)
Detroit Saturday Night (Detroit, Michigan)
Evening Tribune (Fremont, Nebraska)
Fremont Tribune (Fremont, Nebraska)
The Idaho Republication (Blackfoot, Idaho)
Lincoln Sunday Star (Lincoln, Nebraska)
London Standard (London, England)
Humeston New Era (Humeston, Iowa)
Journal (Detroit, Michigan)

Marshall News-Statesman (Marshall, Michigan)
Milwaukee Sentinel (Milwaukee, Wisconsin)
Minneapolis Tribune (Minneapolis, Minnesota)
Minneapolis Sunday Tribune (Minneapolis, Minnesota)
Minneapolis Morning Tribune (Minneapolis, Minnesota)
Moline Daily Dispatch (Moline, Illinois)
New York Times (New York, New York)
The Ogden Standard (Ogden, Utah)
The Omaha Daily Bee (Omaha, Nebraska)
Record (Philadelphia, Pennsylvania)
Times-Republican (Marshalltown, Iowa)
The Topeka Daily State Journal (Topeka, Kansas)
Waterloo-Times Tribune (Waterloo, Iowa)
Waterloo Evening Courier and Reporter (Waterloo, Iowa)
The West Virginian (Fairmont, West Virginia)

Trade Journals and Magazines

Automobile Topics
Commercial Motor
Chilton Tractor Journal
The Commercial Motor
Eastern Dealer
Farm Implement News
Farm Implement News: The Tractor and Truck Review
Farm Collector
Farm Machinery/Farm Power
Ford News
Fortune
The Furrow
Gas Power
The Harvester World
Implement & Tractor Trade Journal
The John Deere Magazine

Motor Age
Scientific American

Articles

Blaney, LeAnne. "When Ford Motors Came to Cork," *Century Ireland*, March 2017.

Ellis, Lynn. "The Problem of the Small Farm Tractor," *Scientific American*, June 7, 1913.

Farm Collector staff. "Iron Age Ads: Converting from Steam to Gas," *Farm Collector*, August 2006.

Hughes, Ralph. "Theo Brown: A Man and His Machines," *John Deere Tradition*, August 2002.

Moore, Sam. "Birth of the Iron Horse: General Motors Attempts a Tractor," *Farm Collector*, October 2003.

Moore, Sam. "Waterloo Boy Tractor Made its Mark in the UK—Under an Alias," *Farm Collector*, April 2018.

Stegh, Leslie. "The Waterloo Strikes of 1919," *The Annals of Iowa*, Third Series, vol. 72, no. 4, Fall 2013.

Vossler, Bill. "On the Trail of the Mysterious Hackney Auto-Plow," *Farm Collector*, April 2000.

Wik, Reynold. "Henry Ford's Tractors and American Agriculture in Agricultural History," *Agricultural History*, vol. 38, no. 2, April 1964, 79–86.

Wik, Reynold. "Nebraska Tractor Shows, 1913–1919, and the Beginning of Power Farming," *Nebraska History 64* (1983): 193–208.

Wise, Charles. "Following the Faultless Engine Co.," *Gas Engine Magazine*, February–March 2016.

Books and Manuscripts

Bak, Richard. *Henry and Edsel: The Creation of the Ford Empire*. Hoboken, New Jersey: John Wiley & Sons, 2003.

Broehl, Wayne. *John Deere's Company: A History of Deere & Company and Its Times*. New York, New York: Doubleday & Company, Inc., 1984.

Bryan, Ford R. *Beyond the Model T: Ford's Other Ventures.* Detroit, Michigan: Wayne State University Press, 1990.

Bowman, J.R., Shearer, Frederick E., eds. *The Pacific Tourist. J.R. Bowman's Illustrated Transcontinental Guide of Travel from the Atlantic to the Pacific Ocean.* New York, New York: J.R. Bowman, 1882.

Brown, Theo. *Deere & Company's Early Tractor Development.* Moline, Illinois: Deere & Company, 1953.

Brown, Theo. *Deere & Company's Early Tractor Development.* Grundy Center, Iowa: Two-Cylinder Club with permission of Deere & Company, 1997.

Collier, Peter, and David Horowitz. *The Fords: An American Epic.* San Francisco, California: Encounter Books, 2002.

Crissley, Forrest. *Alexander Legge, 1866–1933.* Chicago, Illinois: Privately Printed by The Alexander Legge Memorial Committee, 1936.

Dahlstrom, Neil, and Jeremy Dahlstrom. *The John Deere Story: A Biography of Plowmakers John and Charles Deere.* DeKalb, Illinois: Northern Illinois University Press, 2005.

Erb, David, and Eldon Brumbaugh. *Full Steam Ahead: J.I. Case Tractors & Equipment, 1842–1955, Volume 1.* St. Joseph, Michigan: American Society of Agricultural Engineers, 1993.

Federal Trade Commission. *Report on the Agricultural Implement and Machinery Industry.* Washington, DC: Government Printing Office, 1938.

Ford, Henry, with Samuel Crowther. *My Life and Work.* Printed in Lexington, Kentucky, August 24, 2017.

Gardner, Bruce. *American Agriculture in the Twentieth Century: How It Flourished and What It Cost.* Cambridge, Massachusetts: Harvard University Press, 2006.

Gray, R.B. *The Agricultural Tractor, 1855–1950.* St. Joseph, Michigan: American Society of Agricultural Engineers, 1975.

Halberstadt, April. *Case Photographic History.* Osceola, Wisconsin: Motorbooks International, 1995.

Ingram, J.S. *The Centennial Exposition Described and Illustrated.* Philadelphia, Pennsylvania: Hubbard Bros, 1876.

Klancher, Lee. *The Farmall Dynasty: The Story of the Engineering and Design That Created International Harvester Tractors*. Austin, Texas: Octane Press, 2012.

Klancher, Lee. *Tractor: The Heartland Innovation, Ground-Breaking Machines, Midnight Schemes, Secret Garages and Farmyard Geniuses That Mechanized Agriculture*. Austin, Texas: Octane Press, 2018.

Leffingwell, Randy. *John Deere: A History of the Tractor*. St. Paul, Minnesota: Crestline, 2005.

Letourneau, Peter. *John Deere Limited-Production & Experimental Tractors*. Osceola, Wisconsin: Motorbooks International, 1994.

Letourneau, Peter. *John Deere General-Purpose Tractors, 1928–1953*. Osceola, Wisconsin: Motorbooks International, 1993.

Marsh, Barbara. *A Corporate Tragedy: The Agony of International Harvester*. Garden City, New York: Doubleday & Company, Inc., 1985.

McCormick, Cyrus Hall. *The Century of the Reaper*. Boston, Massachusetts: Houghton Mifflin Company, 1931.

Nevins, Allan, and Frank Ernest Hill. *Ford: Expansion and Challenge, 1915–1933*. New York, New York: Charles Scribner's Sons, 1957.

Nolde, Gilbert, ed. *All in a Day's Work: Seventy-Five Years of Caterpillar*. Hong Kong: Forbes Custom Publishing, 2000.

Page, Victor. *The Model T Ford Car, Including the Fordson Farm Tractor, Revised Edition*. New York, New York: The Normal W. Henley Publishing Co., 1922.

Pripps, Robert. *Vintage Ford Tractors*. Stillwater, Minnesota: Town Square Books, 1997.

Pripps, Robert. *Farmall Tractors*. Osceola, Wisconsin: Motorbooks International, 1993.

The Reminiscences of Mr. E.J. Farkas. From the Owen W. Bombard interviews series, 1951–1961. Accession 65 Interview conducted June 1954. Transcript digitized by staff of Benson Ford Research Center, November 2011.

The Reminiscences of Mr. Norman J. Ahrens. From the Owen W. Bombard interviews series, 1951–1961. Accession 65 Interview conducted July 1956. Transcript digitized by staff of Benson Ford Research Center, November 2011.

The Reminiscences of Mr. Howard D. Beebe. From the Owen W. Bombard interviews series, 1951–1961. Accession 65 Interview conducted May 1954. Transcript digitized by staff of Benson Ford Research Center, November 2011.

Renkes, Jim. *The Quad-Cities and The People.* Helena, Montana: American & World Geographic Publishing, 1994.

Sanders, Ralph. *Vintage International Harvester Tractors.* Beverly, Massachusetts: Voyageur Press, 1997.

Schlebecker, John T. *Whereby We Thrive: A History of American Farming, 1607–1972.* Ames, Iowa: Iowa State University Press, 1975.

Seyfarth, A.C. *Tractor History.* McCormick-International Harvester Collection, Wisconsin Historical Society, McCormick Mss 6z, Folder 13864.

Sklovsky Covich, Edith. *Max: Inventor, Engineer, Writer 1877–1967.* Chicago, Illinois: Stuart Brent, Publisher, 1974.

Smith, H.P. *Farm Machinery and Equipment.* New York, New York: McGraw-Hill, 1955.

Sorensen, Charles. *My Forty Years with Ford.* Detroit, Michigan: Wayne State University Press, 2006.

Taylor, W.E. *Soil Cultur*e, Fourth Edition. Moline, Illinois: Deere & Company, 1916.

Taylor, W.E. *Soil Culture and Modern Farm Methods,* Fifth Edition. Moline, Illinois: Deere & Company Soil Culture Department, 1924.

The 1928 Tractor Field Book: Chicago, Illinois: Farm Implement News Co., 1928.

United States Department of Agriculture. *Fourteenth Census of the United States Taken in the Year 1920, Volume 5, Agriculture.* Washington, DC: Washington Government Printing Office, 1922.

United States Department of Commerce. *Commerce Reports, Volume 1, Issue 28, Daily Consular and Trade Reports Issue Daily by the Bureau of Foreign and Domestic Commerce,* Department of Commerce, Washington, DC. Chapter titled "Results Obtained with Tractors in United Kingdom," Consul General Robert P. Skinner, London, December 18, 1917, 443–444.

United States Department of Labor. *U.S. Department of Labor, Bureau of Labor Statistics*, v. 49, July–December 1939, Table 3, Average Income of Farm and

Factory Wage Earners and Indexes of Living Costs, 1909, 1914, and 1919 to 1938.

Williams, Michael. *Massey-Ferguson Tractors*. Alexandria Bay, New York: Reprinted by Farming Press, 1990.

Watts, Steven. *The People's Tycoon: Henry Ford and the American Century*. New York, New York: Alfred A. Knopf, 2005.

Wendell, C.H. *Power in the Past: A History of Gasoline Engine & Tractor Builders in Iowa, 1890–1930*. Hiawatha, Iowa: Wheeler Printing Incorporated, March 1971.

Wik, Reynold. *Henry Ford and Grass-Roots America*. Ann Arbor, Michigan: University of Michigan Press, 1972.

Williams, Robert C. *Fordson, Farmall, and Poppin' Johnny: A History of the Farm Tractor and Its Impact on America*. Urbana, Illinois: University of Illinois Press, 1987.

NOTES

1 / The Ford Tractor

1. Ford, Henry. *My Life and Work*, 9–14.
2. Watts, Steven. *The People's Tycoon: Henry Ford and the American Century*. New York: Alfred A. Knopf, 2005, 14.
3. Exhibits and information on the Centennial Exposition, in Ingram, J. S., "The Centennial Exposition Described and Illustrated." Philadelphia: Hubbard Bros, 1876; Watts, 22.
4. William C. McMillan, investor in the Detroit Dry Dock Company, later invested in Ford.
5. Ford, Henry. *My Life and Work*, 12.
6. Henry Ford cited in Watts, *The People's Tycoon*, 314.
7. Watts, 34, 53; Ford worked at Edison Illuminating Company until 1899, after forming the Detroit Automobile Company with investors on August 5, 1899.
8. Watts, 37.
9. Retelling of Ford's Quadricycle in Watts, 37–52.
10. Ford's early designs and partnerships are in Watts, 85–103.
11. Watts, 85–102.
12. Watts, 102–103.
13. Watts, 102.
14. An average factory worker earned $512 a year. An average farm laborer, for seven months of work, earned half of that. Monthly labor review, U.S. Department of Labor, Bureau of Labor Statistics, v. 49, July–December 1939, Table 3, Average Income of Farm and Factory Wage Earners and Indexes of Living Costs, 1909, 1914, and 1919 to 1938, 64.

15. Watts, 118.
16. Williams, Robert C. *Fordson, Farmall, and Poppin' Johnny: A History of the Farm Tractor and Its Impact on America*. Urbana: University of Illinois Press, 1987, 20.
17. Wik, Reynold. *Henry Ford and Grass-roots America*. Ann Arbor: The University of Michigan Press, 1972, 86.
18. *Farm Implement News*, November 15, 1908.

2 / John Deere: "The Newcomer"

1. Reports on the Haymarket Affair in *The Decatur Republication*, May 5, 1886; *The Wert* (Ohio) *Bulletin*, July 30, 1886.
2. *Monmouth* (Illinois) *Evening Gazette*, May 14, 1884.
3. Marsh, Barbara. *A Corporate Tragedy: The Agony of International Harvester*. Garden City, NY: Doubleday & Company, Inc., 1985, 38–39.
4. Marsh, 40–41.
5. *Oak Park* (Illinois) *Reporter*, September 4, 1902.
6. *Report of the International Harvester Company*, December 31, 1907; *Report of the International Harvester Company*, December 31, 1910.
7. *Report of the International Harvester Company*, December 31, 1907; *Columbia College Campus Preservation Plan*, Volume V, 2005; Harvester moved out of the building at 180 North Michigan Avenue in 1927. Today, the Harvester Building is part of Columbia University's Alexandroff Campus Center. Michigan Avenue construction boom is in *Chicago Examiner*, November 28, 1909.
8. Quote about Moline in Sklovsky Covich, Edith, *Max: Inventor, Engineer, Writer 1877–1967*. Chicago: Stuart Brent, Publisher, 1974.
9. Office description from photograph at Wisconsin Historical Society, International Harvester Company, Office of Cyrus Hall McCormick, Jr., Image 25584.
10. Butterworth description in *Implement & Hardware Journal*, January 5, 1929, 69.
11. *Annual Report of Deere & Company*, 1909; *Report of the International Harvester Company*, December 31, 1909.
12. *Moline Daily Dispatch*, October 29, 1907.
13. "built proportionately" in *Farm Machinery-Farm Power*, No. 1584–1585, June 15, 1922. "tall, awkward . . ." description by Walter Crissley, cited in Ernstes, David P., Hildreth, R. J., and Knutson, Ronald D. *Farm Foundation: 75 years as Catalyst to Agriculture and Rural America*. Oak Brook, IL: Farm Foundation, 2007.
14. Deere & Company Minutes, report of the Advisory Board, October 20, 1909, 187.
15. Broehl, *John Deere's Company: A History of Deere & Company and Its Times*. New York: Doubleday, 1984, 324.
16. Broehl, 324.

17. Cited in Broehl, 343.
18. Deere Plow Company, Mcc Mss 2c BX33, McCormick-International Harvester Collection, Wisconsin Historical Society.
19. John Deere Canadian binder resolutions in Deere & Company Minutes, October 20, 1910, 187–188.
20. Deere & Company Minutes, January 6, 1910, 193.
21. Deere & Company Minutes, Meeting of the Committee on Reorganization, July 1910, 218–219.
22. George Mixter to William Butterworth, August 1, 1923, John Deere Archives, 25603.
23. Joseph Dain obituary in *Farm Implement News*, November 8, 1917.
24. Companies already acquired, or nearly acquired by November, included Union Malleable Iron Company, Kemp and Burpee Mfg. Company, Marseilles Manufacturing Company, Dain Manufacturing Company, Fort Smith Wagon Company, and the Davenport Wagon Company. The Board voted to delay any additional action until "later or after the amalgamation of the already allied interests had been accomplished."
25. Velie in Deere & Company Minutes, May 29, 1912.
26. *International Harvester Annual Report for 1911*, Wisconsin Historical Society; Broehl, 341.
27. "The International World," July 10, 1910, vol. 1, no. 10, 6.

3 / The Tractor Works

1. Johnston biography in *The Harvester World*, December 1922, Wisconsin Historical Society; Seyfarth, Recollections of H.B. Morrow, in *Tractor History*, 120; Seyfarth, Table IV, Gas Tractor Sales. Morrow was later named superintendent of the Tractor Works in late 1915 and held that responsibility for the next five and a half years.
2. E.A. Johnston to A.C. Seyfarth, June 20, 1934.
3. Seyfarth, 30; on the 1908 Winnipeg trials, *Manitoba Free Press*, July 18, 1908; *The Commercial Motor*, August 20, 1908.
4. Seyfarth, 31–32; Seyfarth, Recollection of J.D. McGann, 124.
5. Financials in *International Harvester Annual Report*, 1908; factory shop details in McGann, 124.
6. Seyfarth, 122. A full overview of the Tractor Works was published in *The Iron Age*, vol. 8, no. 26, June 29, 1911, 1582–1586.
7. C.N. Hostester recollection in Seyfarth, 128–129.
8. Ellis, L.W., and Edward Rumely, *Power and the Plow*. New York: Doubleday, 1911, 10.

4 / "Divorce the Plow from the Tractor"

1. Sanders, *Vintage International Harvester Tractors*; slogan for "Mogul 8-16 Tractor Advertising Poster" at Wisconsin Historical Society, Image ID 8764. Overview of early tractor models and where they were built is in Seyfarth, 116C.
2. *Indianapolis News*, August 19, 1908.
3. Deere & Company Minutes, March 5, 1912.
4. W.E. Taylor, *Soil Culture and Modern Farm Methods*, 5th ed.; Deere & Company Soil Culture Department. Moline, Illinois: Deere & Company, 1924, 145; *Prairie Farmer*, September 23, 1858, September 30, 1858, October 21, 1858, November 18, 1858, September 29, 1859. Additional information on the Fawkes steam plow and the Lincoln address is in Gray, *Development of the Agricultural Tractor in the United States*, vol. 1, 3–4.
5. Mansur & Tebbetts Implement Company, Catalog No. 53, 1892.
6. *Implement & Tractor Trade Journal*, August 16, 1916, 23; *Farm Machinery and Equipment*, 26; *Minneapolis Tribune*, December 9, 1906.
7. *The Commercial Motor*, August 20, 1908.
8. *Minneapolis Sunday Tribune*, July 28, 1912.
9. The Winnipeg Agricultural Motor Competition provides a window into the early evolution of the tractor business. Demonstrating a tractor was one thing. But if it didn't pull, push, or move something, it was merely a novelty. In 1908, the Canadian-built Cockshutt was the sole plow used to demonstrate a small slate of tractors, but the following year allowed tractor manufacturers their choice of plow. Deere delivering the largest plow, a 14-inch bottom gang plow. A seven-bottom version was hitched by the Gas Traction Company to its Big Four 30 tractor.
10. John Deere Plow Company, Dallas, Texas, Catalog L, 1912, 160.
11. Diamond Iron Works built tractors for the Transit Threshing Company, among others.
12. Deere & Company Minutes, May 1912, 569–570.
13. Gas Traction sold to Emerson in 1912, *Minneapolis Morning Tribune*, July 18, 1912; biography of Patrick Lyons in *Implement and Tractor Trade Journal*, June 26, 1919, 24.

    P.J. Lyons died of heart disease in 1919 (*Chilton Tractor Journal*, August 1, 1919); Deere & Company Minutes, Executive Committee, March 21, 1912, 522.
14. Negotiations in Deere & Company Minutes, Executive Committee, April 9, 1912, 536–550.
15. Deere & Company Minutes, May 14, 1912; contract with Hart-Parr Co., dated June 28, 1912, John Deere Archives 19022. Quote about gang plow in *Moline Evening Mail and Journal*, November 25, 1912.
16. C.H. Melvin, May 12, 1912, John Deere Archives 41312.

17. Vossler, Bill. "On the Trail of the Mysterious Hackney Auto-Plow," *Farm Collector*, April 2000.
18. February 25, 1914, issue of *Farm Implements* said, "The Hackney Manufacturing Company . . . have sold to the Standard Motor Company of Mason City, Iowa, their entire business, including the good will and patents for the U.S. and Mexico. The real estate and buildings are retained by the present owners. The Hackney interests will be represented in the Standard Motor Company by J.M. and L.S. Hackney, who become members of the board of directors." *Farm Implements* said later, "The Standard Motor Company will continue to manufacture the Hackney Auto-Plow and, in addition, a line of motor trucks and automobiles." Vossler, "On the Trail of the Mysterious Hackney Auto-Plow." Information on staff in *Farm Implements,* March 20, 1912.
19. Melvin's background in *Moline Daily Dispatch*, October 23, 1917; Charles H. Melvin to Deere & Company, May 10, 1912, John Deere Archives 4132; Deere & Company Minutes, July 15 and August 24, 1912. Design of the Melvin tractor in Brown, *Deere & Company's Early Tractor Development*, 3.

5 / "The Great Awakening"

1. Mixter, George. *Report on Implement Trade Conditions in Imperial Russia*, July 14, 1909, John Deere Archives 2356.
2. *Omaha Daily Bee* reported forty tractors on five hundred acres.
3. Wik, Reynold. "Nebraska Tractor Shows, 1913–1919, and the Beginning of Power Farming," *Nebraska History 64* (1983): 193–208; *Omaha Daily Bee*, June 27, 1914.
4. Ellis, Lynn. "The Problem of the Small Farm Tractor," *Scientific American*, June 7, 1913.
5. Watts, 146.
6. *Motor Age*, vol. XXVIII, no. 11, September 9, 1915, 20–21. The "mudhole pirates" story is on page 16. The Yellowstone story is on page 26. Ford sold 500,000 Model Ts in 1915, Willys Overland sold 90,000, and in third place, Dodge sold 45,000.
7. Williams, 24.
8. *Omaha Sunday Bee*, February 22, 1914; "Power Age in Agriculture," in *Implement and Hardware Trade Journal*, vol. XLIV, no. 12, June 8, 1929.
9. International Harvester production data in Seyfarth, 29; McCormick quote in Williams, 24.
10. "Power Age in Agriculture," in *Implement and Hardware Trade Journal*, vol. XLIV, no. 12, June 8, 1929.
11. Toledo *Blade*, May 18, 1915.
12. *The Daily Huronite*, May 18, 1915.
13. *Humeston New Era*, Humeston, IA, May 19, 1915.

14. *Abilene Daily Reporter*, Abilene, TX, May 19, 1915.
15. Model T rebate in Bak, Richard, *Henry and Edsel: The Creation of the Ford Empire.* Hoboken, New Jersey: John Wiley & Sons, 2003, 76.
16. *Dearborn Independent*, October 8, 1915.
17. Watts, 112–114, 315; Wik, 87.
18. Watts, 38, 102; Farkas arrived in New York City from Hungary via the *S.S. Pannovia* in 1906. Equipped with a mechanical engineering degree from the Royal Joseph Technical University in Hungary, he entered the army for a year, then began designing motorcycles with a local firm in his home country. Farkas found work in Brooklyn with a German electrical engineer but was fired when he showed up fifteen minutes late. Three jobs later, he landed with the manufacturer of the Maxwell-Briscoe, a two-cylinder automobile with a double-opposed engine. He spent his time in the drafting room, and not uncommonly, never saw an assembled vehicle and knew nothing of production volumes or the rest of the process. After six months, in 1907, he moved again, sending letters of introduction to Packard and Ford Ferro Engine & Machine Company, among others. Ford Motor Company's Walter Flanders extended an offer of $18 a week, and Farkas found his way to Detroit.
19. *The Reminiscences of Mr. E.J. Farkas,* from the Owen W. Bombard interviews series, 1951–1961 Accession 65 Interview conducted June 1954. Transcript digitized by staff of Benson Ford Research Center: November 2011, 46, https://cdm15889.contentdm.oclc.org/digital/collection/p15889coll2/id/7759.
20. Watts, 114; *The Reminiscences of Mr. E.J. Farkas,* 70.

6 / "Small Tractor Proposition"

1. Agricultural Census, 1920 (Chapter 1, Farms and Farm Property); 2017 figure from USDA, http://usda.mannlib.cornell.edu/usda/current/FarmLandIn/FarmLandIn-02-16-2018.pdf. Agricultural historian Bruce L. Gardner has charted the number of horses and tractors on farms in his book *American Agriculture in the Twentieth Century.* Farm size includes "unimproved" land, timber, or other land that could not be planted.
2. "Why and How the Farm Boy Will Beat Dad's Time," a radio address broadcast from KYW September 29 by Walter B. Remley, Agricultural Extension Department, International Harvester, printed in *Farm Implement News*, October 8, 1925, vol. 46, no. 41.
3. Advertisement in *The Daily Gate City*, Keokuk, Iowa, August 26, 1910.

4. *Daily Dispatch,* April 13, 1937; Max Sklovsky Biographical File, John Deere Archives 26059; Sklovsky Covich, Edith, *Max: Inventor, Engineer, Writer 1877–1967* (Chicago: Stuart Brent, Publisher, 1974). Darwin quote in *Max,* 11–12. Childhood of Pauline in *Max,* 20–24; letters to Pauline, including early descriptions of Moline in *Max,* 42–46; "got in with good people . . ." in Max, 73. Brown Diary, July 19, 1915; details on ammunition carts in Max, 79–81; on trip to Germany in *Max,* 82.
5. *Report on the Manufacturing Investment of Deere & Company for the Year Ending October 31, 1914,* John Deere Archives 2698.
6. George Mixter to William Butterworth, May 17, 1915, John Deere Archives 4132.
7. Deere & Company Minutes, March 14, 1916.
8. Brown diary, February 16, 1916.
9. Deere & Company Minutes, June 13, 14, 1916, pp. 1960–1985.
10. Deere & Company Minutes, July 13, 1916. This is the first reference to the "necessity" of kerosene in Deere Minutes or correspondence.

7 / Henry Ford Day

1. Bowman, J.R., Shearer, Frederick E., ed. *The Pacific Tourist. J.R. Bowman's Illustrated Transcontinental Guide of Travel from the Atlantic to the Pacific Ocean.* New York: J.R. Bowman, 1882, 23. OCLC 752667534.
2. Description of Fremont in *Omaha Daily Bee,* August 3, 1916.
3. *Lincoln Sunday Star,* August 13, 1916, 11.
4. *Fremont Tribune,* August 2, 1917.
5. "The National Tractor Demonstrations," *Gas Power,* July 1916, 55–58.
6. *Gas Power,* July 1916, 58.
7. *Farm Implement News,* July 27, 1916, Waterloo Boy ad lists the circuit, page 4.
8. McCormick accident in *Chicago Examiner,* April 27, 1916.
9. *The Bee,* Omaha, August 10, 1916.
10. *Special Omaha-Council Bluffs Edition of the Implement & Tractor Trade Journal,* August 1916.
11. Wisconsin Historical Society, McCormick Collection, advertising poster for International Harvester kerosene tractor demonstration at Cedar Rapids, Iowa. Features color illustrations of the Mogul 8-16, Titan 10-20, Titan 15-30, Titan 30-60, and Mogul 12-25 tractors. Includes color illustration of tractors. Image 4282.

12. Tents advertised in the *Omaha Daily Bee*, August 3, 1916; Wallie's tree-climbing escape in *Omaha Daily Bee*, August 10, 1916.
13. Wik, 88, cites Topeka *Daily Capital*.
14. Fremont *Evening Tribune*, August 7, 1916; Wolz in the *Omaha Daily Bee*, August 3, 1916.
15. Wik, 88 cites *Daily Bee*, August 8, 1916.
16. *Farm Implement News*, August 17, 1916, 19–20.
17. Brown diary, August 9, 1916.
18. Seyfarth, 43.
19. *Gas Power*, October 1916, 37–38.
20. *Farm Implement News*, August 17, 1916, 19–20.

8 / "A War to End All Wars"

1. *The Alma Record*, July 29, 1915; "Purchase Paves Way for New Ford Era," *Motor Age*, July 17, 1919, 10–11; "Ford Buys Thousand-Acre Site for Blast Furnaces and Plant," *Automobile Topics*, June 19, 1915, 435; Detroit *Journal*, June 16, 1915; Edsel Ford to dealers, March 21, 1919, accession 78, box 1; Grand sub-division advertisement in *Detroit Times*, October 15, 1915.
2. *Marshall News-Statesman*, September 30, 1916.
3. Watts cites the *Detroit Saturday Night*, January 28, 1928, 2.
4. Watts, 30.
5. *The London Times*, September 28, 1915.
6. *Detroit Free Press*, August 22, 1915; Ford quote on Wilson in Watts, 228.
7. *Motor Age*, vol. XXVIII, no. 11, September 9, 1915, 14.
8. Philadelphia *Record* cited in Watts, 231.
9. Watts, 235.
10. *The London Times*, September 28, 1915, cited in Wik, 89.
11. Ford, *My Life and Work*, 95.
12. *The Furrow*, March–April, 1904, vol. IX, no. 2; Blaney, Dr. Leann. "When Ford Motors Came to Cork: The Story of the American Motor Giant Who, in a Time of War and Revolution, Opened a Manufacturing Plant on the Banks of the River Lee." *Century Ireland*, March 2017.
13. Ford Fair Lane Papers, telegram, Percival Perry, London, to Edsel Ford, Dearborn, Michigan, April 7, 1917, Accession 62, Box 519.
14. Estimate of three thousand Overtime tractors in *The Commercial Motor*, November 27, 1917. Reports on American tractors in "England Wants More Tractors," *Automotive Industries*, vol. 37, November 22, 1917, 920–921.
15. *The Commercial Motor*, April 2, 1917.
16. *The Commercial Motor*, August 2, 1917.

17. *The West Virginian*, Fairmont, July 26, 1917.
18. Lord Northcliffe to Henry Ford, October 9, 1917, in Nevins, 61.
19. Ford, *My Life and Work*, 95.
20. Ford, *My Life and Work*, 97.

9 / "First Class All the Way"

1. *Farm Implement News*, September 24, 1916, 26; *Farm Implement News,* September 21, 1916.
2. Financial page of the *Boston Sunday Post*, December 31, 1916; the Ford Tractor Company, Inc., issued its first public offering for $10 million in treasury shares in late 1916; *Gas Power*, September 1916, 70.
3. Williams, 70. Gray, *The Agricultural Tractor*, 60. Test number one took place from March 31 to April 9, 1920, a John Deere Waterloo Boy "N", serial number 19851, rated at 12–25 horsepower. Chase's involvement outlined in Wendel, 8.
4. Deere & Company Minutes, November 1916. At the time, they were not overly impressed with the Waterloo Boy. "We do not believe that the Waterloo Boy tractor, except under very favorable conditions, can pull more than one 16-inch sulky plow and stand up."
5. Motor Cultivators at the 1920 show in Wichita from *Implement and Tractor Trade Journal*, vol. XXXIIV, no. 28, July 12, 1919, 58.
6. *Dubuque Democratic Herald*, July 5, 1865; advertisements for Hawkeye Cultivator, 1863 and 1867, in John Deere Archives 19207, 26052.
7. Henry Ford & Son capitalization in Bryan, 22; "A Day with the Doctor," *Eastern Dealer*, April 19, 1917, in W.E. Taylor Memorial Scrapbook, John Deere Archives AN2006-0045-12.
8. In March 1917, George Mixter updated the Board on the full tractor development program.
9. W.E. Taylor, *Soil Culture*, 4th ed., Moline, IL: Deere & Company, 1916, 41.
10. Deere & Company Minutes, March 14, 1917, 2151–2152.
11. Deere & Company Minutes, June 12, 1917, 2233–2235.
12. Deere & Company Minutes, September 12, 1917, 2302–2303.

10 / "England Gets Them First"

1. "Little Americans Do Your Bit" poster, Wisconsin Historical Society, Image ID 71215.
2. Constitution and By-Laws of the National Association of Agricultural Implement and Vehicle Manufacturers of the United States of America, Approved October 11, 1894, John Deere Archives 136002; Minutes, Executive Committee

Meeting of the National Implement and Vehicle Association; Chamber of Commerce notes are page 172–173; meeting with editors is page 184; discussion of National Association of Tractor and Thresher Manufacturers is on page 176 (April 12, 1917) and page 194; others are minutes from September 13, 1917, 197; Wolverine drop on May 15, 1919, 296, John Deere Archives 136008.

3. *The West Virginian*, Fairmont, July 26, 1917.
4. *Commerce Reports, Volume 1, Issue 28, Daily Consular and Trade Reports Issue Daily by the Bureau of Foreign and Domestic Commerce,* Department of Commerce, Washington, DC, chapter titled "Results Obtained with Tractors in United Kingdom," Consul General Robert P. Skinner, London, December 18, 1917, 443–444; list of entries in booklet for the International Tractor Trials, and Agricultural Machinery Exhibit 1919 (24–27 Sep) at the Museum of English Rural Life, University of Reading, General Show and Exhibition Records series TR 2RAN/SH3 TR 2RAN/SH3/3/8; "Organised by the Highland & Agricultural Society of Scotland, at East Graigie Home farm, Crammond Bridge, Edinburgh, at Blackhill Farm, Mayhill, Glasgow, and at Spoutwells and New Mains Farms, Scone Perth. Tractors were manufactured by: Alldays & Onions Pneumatic Engineering Co. Ltd.; Bullock Tractor Co.; Wallis Tractor Co.; W. Weeks & Son Ltd.; Samson, Sieve-Grip Tractor Co.; Allis-Chalmers Manufacturing Co.; Emerson Brantingham Implement Co.; Cleveland Motor Tractor Co.; Chase Tractor Co.; J.I. Case Threshing Machine Co.; John Fowler & Co. (Leeds) Ltd.; Gaston, Williams & Wigmore Ltd.; James Hodgson; International Harvester Co.; Mann's Patent Steam Cart & Waggon Co. Ltd.; Parrett Tractor Co.; Martin's Cultivator Co. Ltd.; E.G. Staude Manufacturing Co.; Saunderson Tractor & Implement Co.; Overtime Farm Tractor Co.; Joliet Oil Tractor Co.; Killen-Strait Manufacturing Co.; Moline Plow Co.; Bull Tractor Co.; Wyles Motor Ploughs Ltd." The "Scottish Tractor Trials" were reported in *The Commercial Motor*, October 25, 1917.
5. Wik, Reynold. "Henry Ford's Tractors and American Agriculture in Agricultural History," 79–86; Pripps, Robert N. *Vintage Ford Tractors.* Jackson, MS: Town Square Books, 1997; Ford's claim in *My Life and Work*, 97; Bryan, *Beyond the Model T,* rev. ed. Detroit, MI: Wayne State University Press, 1997, 21.
6. Ford, *My Life and Work.* Charles Sorensen, *My Forty Years with Ford*, 234.
7. ". . . all motor and no frame" quoted in Wik, 92.
8. Bryan, "Beyond the Model T," 21–23.
9. *The Ogden Standard*, June 15, 1918.
10. *The West Virginian*, July 26, 1917.
11. Gay, Larry. *A Guide to Ford, Fordson, and New Holland Tractors, 1907–1999; The Idaho Republication*, July 12, 1918.

11 / The John Deere Tractor

1. Deere & Company Minutes, September 12, 1917; ibid, November 7, 1917.
2. *Ottumwa Semi-Weekly Courier*, November 2, 1917.
3. Clausen, Leon R., Biographical Data, John Deere Archives 24938.
4. Deere & Company Minutes, November 19, 1917, 2337–2344; *Report on the Manufacturing Investment of Deere & Company for the Year Ending October 31, 1917*, John Deere Archives 2704.
5. Willard Velie to Burton Peek, January 15, 1918, John Deere Archives 24487.
6. Dineen to Mixter, February 21, 1918, George Peek Papers, Missouri Historical Society.
7. Ford's reference to J.D. Oliver in Brown, April 22, 1918.
8. George Peek to Henry Ford, February 26, 1918, George Peek Papers, Missouri Historical Society.
9. H.B. Dineen to George Peek, George Peek Papers, Missouri Historical Society; completion of "14 experimental plow for Fordson tractor" in Brown, February 15, 1918.
10. Patent no. 550,266 was granted in 1895.
11. According to an article in the Dubuque (Iowa) *Evening Times*, a steam traction engine cost $15 a day to operate, while the Froelich engine claimed to run on $7 a day.
12. E.B. Parkhurst created a variety of tractor and engine designs. He moved to Waterloo in 1911 from Moline and was killed in 1915. He was driving a new tractor, of his own design, across the Union Pacific railroad tracks in Cheyenne, Wyoming, when a train came around a bend and struck him. He was killed instantaneously. (*Moline Daily Dispatch*, August 23, 1911, on his move to Waterloo; April 20, 1915, on his death).
13. Deere & Company Minutes, December 12, 1916, 2087–2092.
14. William Butterworth to Burton Peek, September 8, 1916, John Deere Archives 24487; United States Agricultural Statistics Board, Agricultural Marketing Service, United States Bureau of Agricultural Economics, United States Crop Reporting Board. *Agricultural Prices.* Washington, DC: Crop Reporting Board, 1942.
15. *Waterloo-Times Tribune*, March 24, 1918; onmilwaukee.com, Borchert trial took many twists and turns before reaching a conclusion, August 14, 2011. Borchert used his proceeds from the deal, $300,000 total, and bought an interest in the Milwaukee Brewers baseball club of the American Association. In 1927, standing at a podium in front of 700 baseball fans at a dinner held at the Elks Club, not to mention a live radio audience, "death called three strikes" as Borchert suffered a stroke. *Milwaukee Sentinel*, April 29, 1927.

16. Deere & Company Minutes, January 25, 1918. Otto Borchert to Isabell Borchert, incorrectly dated 1917, Milwaukee Historical Society.
17. Sears, Roebuck & Company, Balance Sheet and Statement for Fiscal Year Ending December 31, 1918, http://www.library.upenn.edu/collections/lippincott/corprpts/sears/sears1918.pdf); farm implements in Sears, Roebuck and Company Catalog, 1918, https://archive.org/details/catalog1918sear_201901.
18. Louis Witry biographical file, John Deere Archives, 22615.
19. *Waterloo Evening Courier and Reporter*, December 31, 1917; *Times-Republican*, Marshalltown, Iowa, March 23, 1918; Deere & Company Minutes, March 12, 1918.
20. *Boston News Bureau*, November 18, 1918, in William Butterworth Scrapbook F, John Deere Archives 2006-0043-11.
21. Deere & Company Minutes, March 12, 1918.
22. Charles Deere Velie to William Butterworth, March 30, 1918, George Peek Papers, Missouri Historical Society.
23. *John Deere Magazine*, May 1918, 2–4.
24. "Waterloo Gasoline Traction Engine Co.: Origin & History," John Deere Archives 11818; "Report on the Manufacturing Investment of Deere & Company for the Fiscal Year Ending October 31st, 1921," John Deere Archives, 45190.
25. Waterloo Boy description and testimonial from "Waterloo Boy Tractor: The Original Kerosene Tractor," Series A-67-Waterloo-1919," 1919, John Deere Archives 31471.
26. "Tractor factory" reference in 1920 Census, description of All-Wheel Drive in Maryland in Martin Fleming to Theo Brown, April 19, 1945, John Deere Archives 1123.

12 / "Ford Likes a Success"

1. McCormick, Cyrus Hall. *The Century of the Reaper*. Boston: Houghton Mifflin Company, 1931, 162–163.
2. Minutes, Executive Committee Meeting of the National Implement and Vehicle Association, May 16, 1918, 240–244, John Deere Archives 136008.
3. Brown, April 22, 1918.
4. "Tractor Drawn Implements," April 29, 1918, George Peek Papers, Missouri Historical Society.
5. Telegraph, Harold Dineen to George Peek, March 4, 1918, George Peek Papers, Missouri Historical Society.
6. Harold Dineen to George Peek, March 30, 1918, George Peek Papers, Missouri Historical Society; Oliver plow use in the UK in *The Commercial Motor*, "The Oliver No. 7 Plough, Its Use and Misuse," September 5, 1918.

7. "Tractor Drawn Implements," minutes of meeting held April 29, 1918, George Peek Papers, Missouri Historical Society.
8. Watts, 114–115.
9. Bak, *Henry and Edsel: The Creation of the Ford Empire*, 119.
10. Deere implement development and tests at Ford farm in "Tractor Drawn Implements," April 29, 1918, George Peek Papers, Missouri Historical Society; Floyd Todd to George Peek, June 22, 1918; Floyd Todd to E.R. Bryant, June 22, 1918; E.R. Bryant to Floyd Todd, June 18, 1918, George Peek Papers, Missouri Historical Society.
11. Ford's distribution discussions in Brown, May 24, 1918, July 16, 1918, July 17, 1918.
12. Theo Brown Diaries, Worcester Polytechnic Institute, entries February 15–October 5, 1918.
13. Stephen Velie report quoted in Broehl, 409.
14. Deere & Company Blue Bulletin 186, April 4, 1918.
15. Deere & Company Bulletin 220, July 23, 1918.
16. Deere & Company Minutes, September 3, 1919, 2556; Deere & Company Bulletin 229, September 23, 1918.
17. Deere & Company Blue Bulletin 220, July 23, 1918.
18. Harold Dineen, "Report of Visit to Henry Ford & Son on July 16th and 17th [1918]," George Peek Papers, Missouri Historical Society.

13 / Tractor City

1. *Farm Implement News*, September 21, 1916, 24.
2. *The Harvester World*, vol. 8, no. 9, September 1917, 14; *The Harvester World*, July 1918.
3. *Abilene* (Kansas) *Weekly Reflector*, July 25, 1918.
4. *The Topeka Daily State Journal*, July 29, 1918.
5. *Implement and Tractor Trade Journal*, vol. XXXIII, no. 32, August 10, 1918.
6. Advertisement in *The Topeka State Journal*, July 17, 1918.
7. *Implement and Tractor Trade Journal*, vol. XXXIII, no. 32, August 10, 1918, 42-j, 42-k; *The Century of the Reaper*, 185–186. McCormick reported the dealers of 21,800 in 1917 was reduced to 13,860 by 1919.
8. Advertisement in *Farm Implement News: The Tractor and Truck Review*, vol. XXXIX, no. 36, September 5, 1918, 15. Hession was based at 27 Hewett Street, Buffalo, New York.
9. *Farm Implement News: The Tractor and Truck Review*, vol. XXXIX, no. 36, September 5, 1918, 19; information on Massey Harris relationship in Michael Williams, *Massey-Ferguson Tractors*, 15–18.

10. From interview with F. Lee Norton, Racine, Wisconsin, December 28, 1946, cited in Wik, *Henry Ford and Grass-Roots America*, 98; J.I. Case Threshing Machine Co. ad in *System on the Farm*, vol. 6, no. 2, February 1920.
11. Reference to Wallis tractor owners in *Implement and Tractor Trade Journal*, vol. XXXIII, no. 32, August 10, 1918, 42-e. Wallis advertisement in *Implement and Tractor Trade Journal*, vol. XXXIII, no. 33, August 17, 1918, 31. Selby in *Abilene Weekly Reflector*, August 8, 1918.
12. Meetings recounted in "Report of Conference, F.R. Todd and Theo Brown with Henry Ford and C.E. Sorensen," 6 November 1919, John Deere Archives 46548; "Theo Brown Diaries, 29 and 30 July 1918," 2 August 1918.
13. For Salina show, see *Farm Implement News: The Tractor and Truck Review*, vol. XXXIX, no. 32, August 8, 1918, 42–50. Oliver and Grand Detour Plow Company tie-up in Broehl, 410. Oliver representation at Salina, see Charles Deere Velie to George Mixer, August 5, 1918, George Peek Papers, Missouri Historical Society.
14. *Implement and Tractor Trade Journal*, vol. XXXIII, no. 32, August 10, 1918, 4–5.
15. Broehl, 412.
16. *Davenport Daily Democrat*, July 14, 1917; according to the *Moline Daily Dispatch*, March 15, 1917, the tractor debuted in Kansas City, and Velie purchased fifteen acres in Moline. Velie Biltwell Tractor advertisement, Wisconsin Historical Society, MCC MSS 6Z, Box 1052, Folder 44.
17. *New York Times*, August 16, 1918.
18. Charles Deere Velie to George Mixer, August 5, 1918, George Peek Papers, Missouri Historical Society.
19. Deere & Company Blue Bulletin No. 229, September 23, 1918.
20. George Peek to Harold Dineen, March 30, 1918, George Peek Papers, Missouri Historical Society.
21. Brown, September 12, 1918.

14 / "Better, Cheaper"

1. *Moline Daily Dispatch*, October 21, 1918.
2. Benjamin Holt and C.L. Best merged to form the Caterpillar Tractor Company in 1925. Nolde, Gilbert, ed. *All in a Day's Work: Seventy-Five Years of Caterpillar*. Hong Kong: Forbes Custom Publishing, 2000.
3. *Automobile Topics*, March 1919, quoted in Nevins, 147.
4. Velie contracts detailed in *Motor Age*, vol. XXXV, no. 2, January 9, 1919, 14.
5. *Motor Age*, October 10, 1918.

6. Pripps, Robert. *Farmall Tractors*. Osceola, WI: Motorbooks International, 1993, 13.
7. *Farm Implement News*, March 13, 1919. There is no discussion of a Ford acquisition in the Deere & Company Board Minutes.
8. Lawsuit in Nevins, 89–91 and Watts 254–258, 276.
9. Sorensen, *My Forty Years with Ford*, 49.
10. *Los Angeles Examiner*, March 5, 1919; *Los Angeles Sunday Times*, March 16, 1919.
11. Dodge statement quoted in Watts, 278.
12. Watts, 281; Nevins 108–111.
13. Friend comments in *Implement & Tractor Trade Journal*, vol. 34, no. 34, August 23, 1919; Railsback in *Implement & Tractor Trade Journal*, vol. 34, no. 31, August 2, 1919.
14. Salina coverage in *Implement and Tractor Trade Journal*, vol. XXXIII, no. 32, August 10, 1918.
15. According to the *John Deere Magazine*, April 1919, 14, 80,000 people attended the National Tractor Show in Kansas City from February 10–15, 1919. "Display of the Waterloo Boy Tractor was in charge of W.H. Oliver, who was assisted by many travelers and factory men. Prospects were received from all parts of the United States and Canada. Moline men who attended were Messrs. B.F. Peek, H.B. Dineen, Theo. Brown, C.H. Gamble, F.J. Sprung, H.M. Railsback, R.E. Swartley, and Frank Raisbeck, and R.C. Livesay of the John Deere Harvester Works; advertisement "Now for the Greatest of All! Fourth Annual National Tractor Show" in *Implement and Tractor Trade Journal*, vol. XXXIII, no. 32, August 10, 1918, 4–5.
16. Seyfarth, 50.
17. *Implement & Tractor Trade Journal*, vol. XXXIV, no. 10, March 8, 1919, 37. Preston's enlistment in *Implement and Tractor Trade Journal*, vol. XXXIII, no. 33, August 17, 1918, 18.
18. *Implement and Tractor Trade Journal*, vol. XXXIII, no. 36, September 7, 1918, 14.
19. Pricing in *Implement and Tractor Trade Journal*, vol. XXXIV, no. 30, July 26, 1919, 24.
20. Meetings recounted in "Report of Conference, F.R. Todd and Theo Brown with Henry Ford and C.E. Sorensen," 6 November 1919, John Deere Archives 46548.
21. Ford's lawsuit against the *Chicago Tribune* is outlined in Watts, 266–270. "man with a vision" . . . from *Sioux City* (Iowa) *Journal* cited in Watts, 269; *The Nation* cited in Watts, 269.
22. Brown entry, November 6, 1919; "No. 40 and No. 45 Plows," Deere & Company Bulletin No. 333, November 15, 1919; "Fordson Sales Policy," Deere & Company Bulletin No. 382, May 25, 1920.

15 / "Depression Is Awful"

1. *Implement and Tractor Trade Journal*, vol. XXXVI, no. 3, August 2, 1919.
2. Seyfarth, 51, 90; John N. Willys acquired the Moline Plow Company from the owners, the Stephens family, and with it the Universal tractor and other agricultural implements. Advertisement for the Universal in the *Moline Daily Dispatch*, December 9, 1916.
3. *The Clovis News*, June 14, 1917.
4. Progress of General Motors tractor business can be found in Report of General Motors Corporation for the years ending December 31, 1917, December 31, 1918, December 31, 1919, and December 31, 1922. Sam Moore, "Birth of the Iron Horse: General Motors Attempts a Tractor," *Farm Collector*, October 2003. In the 1922 Annual Report, GM reported an initial investment of $10,428,416, an increase of $3,021,034 "principally to cover overrun expenditures" and $7,000,000 in allotments for inventories in May 1920.
5. Quoted in Williams, 97.
6. Indian advertisement in the *Kansas Farmer*, January 9, 1915.
7. Taylor, *Soil Culture and Modern Farm Methods*, 5th ed., 147.
8. Pest description in Taylor, 222–228; W.R. Walton cited in Taylor, 232.
9. Information on growing seasons and methods from Leavens, George D. "Corn: The Foundation of Profitable Farming." New York: The Coe-Mortimer Company, a subsidiary of The American Agricultural Chemical Co., 1915. Leavans was formerly of the Department of Fertilizers, Massachusetts Agricultural Experiment Station. Corn variety names in Taylor, 200, 235.
10. Taylor, 200–201.
11. *Implement and Tractor Trade Journal*, vol. XXXIII, no. 33, August 17, 1918, 32.
12. Waterloo Gasoline Engine Company Decision No. 436, effective January 20, 1920, John Deere Archives 5869.
13. Deere & Company Minutes, May 27, 1920, 3001.
14. Gay, 6.
15. The demand for increased production during the war drove enormous gains in the sale of farm implements. Average farm income grew from $107 million to $213 million during the course of World War I, and increased to $285 million in 1921. Federal Trade Commission Report on the Agricultural Implement and Machinery Industry, 1938, 166.
16. Alexander Legge at meeting of American Farm Bureau Federation, reported in *Lawrence County News*, November 1, 1922; Williams, 72.
17. *Farm Implement News: The Tractor and Truck Review*, September 28, 1922, 12.

18. Legge conversation in "The Century of the Reaper," 197–200; article by Charles L. Hays, "Buy More Fords Than Implements," in John Deere Archives Scrapbook; C. Ford quote in Watts, 97.
19. Nevins, 152–155.
20. Deere & Company, March 21, 1921, in John Deere Archives, Scrapbook C, 1921–1924.
21. Willard Velie to William Butterworth and T.F. Wharton, in Deere & Company Minutes, April 26, 1921.
22. Deere & Company Branch House Bulletin No. 431, January 29, 1921.
23. Brown diary, April 6, 1921; Deere & Company Branch House Bulletin No. 459, August 27, 1921.
24. Deere & Company Minutes, October 18, 1921.
25. Deere & Company Annual Report, 1921, 1922.
26. *New York Times*, February 14, 1922.
27. Broehl, 448. Ford sold 67,000 tractors in 1920, 35,000 in 1921, 67,000 in 1922, and 101,898 in 1923.
28. Leon Clausen, John Deere Archives, 12436.
29. Alex Legge, "The Implement Industry and Prospects for 1924," *Farm Implement News Tractor and Truck Review*, January 24, 1924, 11.

16 / Farmall

1. Only an executive named J.F. Jones opposed the project.
2. Seyfarth, 89–90.
3. Seyfarth, 93.

17 / Iron Man

1. John Deere Archives, Scrapbook C, 1921–1924. Harriet Wardman McCormick obituary, "Mrs. Cyrus Hall McCormick Dies," in *Moline Daily Dispatch*, January 17, 1921.
2. "built proportionately" in *Farm Machinery-Farm Power*, No. 1584–1585, June 15, 1922. Farm Foundation 75th; "tall, awkward . . ." description by Walter Crissley, cited in *Farm Foundation: 75 Years as Catalyst to Agriculture and Rural America*, 8.
3. Typewritten response by William Erwin to Farm Foundation survey, January 21, 2003, Archives of Farm Foundation, 1301 West 22nd Street, Suite 615, Oak Brook, IL, cited in *Farm Foundation: 75 Years as Catalyst to Agriculture and Rural America*, 144; "There's your man . . ." quote in Crissley, 127.

4. Letter from R.J. Hildreth, Farm Foundation, to James W. Cook, Cook, Nelson and Tuthill, Inc., January 22, 1973, Archives of Farm Foundation, 1301 West 22nd Street, Suite 615, Oak Brook, IL, cited in *Farm Foundation: 75 Years as Catalyst to Agriculture and Rural America*, 145–146.
5. Legge biographical information in *Farm Foundation: 75 Years as Catalyst to Agriculture and Rural America*; farmer income fell to $3.9 billion in 1921 from $9.6 billion in 1919, according to *Farm Foundation: 75 Years as Catalyst to Agriculture and Rural America*.
6. Advertising for Harvester's installment plan in *The River Press* (Fort Benton, MT), June 1, 1921.
7. "owners can do their work . . ." advertisement in the *Brownston* (Indiana) *Banner*, December 13, 1922.
8. *Tractor Men Out to Compete with Market for Staple Farm Products*, leaflet published by the Horse Association of America, 1922, in Scrapbook C, 1921–1924, John Deere Archives.
9. *Chicago Journal of Commerce*, February 4, 1922, in Scrapbook C, 1921–1924, John Deere Archives.
10. Uncited article, by J.L. Jenkins, dated 1922, in John Deere Archives Scrapbook C, 1921–1924; a scrapbook kept at Deere's headquarters features the daily coverage of the falling ag markets and the price wars that ensued. An article is pasted into the book, with the bold headline "Ford Hopes to Put a Tractor on Every Farm: Million a Year Is His Goal."
11. Ford, *My Life and Work*, 99.
12. *The Osgood Journal*, Osgood, IN, February 22, 1922.

18 / "Our Main Competition"

1. Deere & Company Minutes, February 22, 1922, 3367–3368.
2. Waterloo sales in Comptroller's Report, 1922; All-Wheel Drive and Waterloo Boy price in January 15, 1923, contract, Deere & Webber Co., John Deere Archives 126247; Deere & Company Minutes, April 1922.
3. Progress of General Motors tractor business can be found in Report of General Motors Corporation for the Year ending December 31, 1917, December 31, 1918, December 31, 1919, and December 31, 1922. Total losses amounted to $33,240,044. Sam Moore, "Birth of the Iron Horse: General Motors Attempts a Tractor," October 2003. In the 1922 Annual Report, GM reported an initial investment of $10,428,416, an increase of $3,021,034 "principally to cover overrun expenditures" and $7,000,000 in allotments for inventories in May 1920.

4. Shareholder Frank Allen remained on board as part of the transition plan, but not for long. William Van Dervoort, formerly the chief supplier of stationary engines for Deere before transitioning to build engines for the Moline Plow Company automobiles, died late in 1921. The problems continued to mount. *Moline Daily Dispatch*, August 13, 1919; Harold Dineen's transition in *Moline Daily Dispatch*, September 25, 1919. Dineen was named assistant to Vice President Robert Lea.
5. *Moline Cultivator for the Fordson*, RF242-10-26M, John Deere Archives.
6. *Moline Daily Dispatch*, June 13, 1921.
7. Gray, vols. 2, 3. The United States Tractor & Machine Company changed its name to the Wisconsin Automobile Company in 1923.

19 / "Stick to It"

1. Clausen, John Deere Archives, 12436. International Harvester price cut from Leffingwell, *John Deere: A History of the Tractor*, 86.
2. Seyfarth, 56.
3. Seyfarth, 96.
4. "one man outfit" in Seyfarth, 116E, entry for 1916. The Farmall replacing the horse in Seyfarth, 116F, entry for 1920.
5. *The* (Chicago) *Day Book*, April 12, 1915; *Chicago Sunday Tribune*, March 21, 1926, included in Deere & Company Scrapbook D. Insull farm information at "Hawthorn-Mellody Farms & Amusement Park, Libertyville, IL. (1907–1970)," *Digital Research Library of Illinois History Journal* (blog), April 18, 2017.
6. Farmall development history in Seyfarth, 91–117.
7. Legge's statement is in *Oxnard* (California) *Daily Courier*, July 20, 1923.
8. Comment on Ford in Seyfarth, 105; McCormick-Deering description in *Greensburg Daily News*, August 28, 1923.
9. Everson views in Seyfarth, 100–101; Legge to E.A. Johnston, April 27, 1923, cited in Seyfarth, 101–102.
10. Development of the Farmall tractor in Seyfarth. Early experimental models in Fiodin Report in Seyfarth, 102. Farmall tractor numbers started at QC-500. The official Farmall serial number record does not include the first 200 (except QC-501, built February 29, 1924. The next entry is QC-701, built December 1924. A total of 5,468 Farmall tractors were built at the Chicago Tractor Works. The first from the Farmall Works in Rock Island, T-5969, was built in 1926. Serial Number List, Farmall Works, 1926–1971, Wisconsin Historical Society, McCormick Collection.
11. Williams, 87–88.

20 / "A Vengeance"

1. In 1922, 36,753 Fordson tractors were built.
2. Articles of adoption come from the *Ford News*. First half statistics for 1923 in vol. III, no. 18, July 15, 1923; *Ford News*, vol. II, F, July 22, 1923; *Ford News*, vol. III, no. 19, August 1, 1923; *Ford News*, vol. II, no. 2, July 22, 1923; *Ford News*, December 22, 1922; *Oakland Tribune*, July 29, 1923.
3. Nevins, Allan, *Ford: Expansion and Challenge*, 366–370.
4. Worldwide operations in Nevins, 255, 370–372, Watts, 345.
5. Seyfarth, 57.
6. Bak, Richard. *The Creation of the Ford Empire: Henry and Edsel.* Hoboken, NJ: John Wiley & Sons, Inc., 2003, 131–134. Fordson sales dropped to 83,000 units in 1924; record sales exceeding 104,000 closed out 1925, then dropped to 88,000 in 1926.
7. Kanzler's memorandum from 1926 in Collier and Horowitz, 95.

21 / "High Hopes"

1. Deere & Company Minutes, April 29, 1924.
2. *Daily Times*, February 29, 1924; full text of address in *Moline Dispatch*, February 29, 1924. Both in Scrapbook G, 2006-0041-11, John Deere Archives.
3. *Farm Implement News*, vol. 46, no. 34, August 20, 1925, 14, based on report of the Department of Commerce.
4. *Farm Implement News*, "A Survey of Equipment on Illinois Farms," vol. 46, no. 40, October 1, 1925, 18.
5. John Deere Dain Tractor price drop recorded in Deere & Webber Co. January 15, 1923 Contract, John Deere Archives 126247. The insert listed the net price (dealer) at $637.50, and list price at $750.
6. Wheat yield per acre is in Taylor, 248–249. Elevations and soil conditions are in Taylor, 251.
7. "Experimental Models Made Up Preceding Adoption and Production of Model 'D' Tractors," 1954, John Deere Archives, 1127.
8. Broehl, 450.
9. H.G. Glessner to Waterloo Gasoline Engine Company, June 18, 1923, John Deere Archives, 1127.
10. Talk made by H.M. Railsback at Waterloo, Iowa, at a meeting of John Deere Branch House and Factory Managers and Others, June 28, 29, 1938, John Deere Archives, 40864, "Speeches."
11. Deere & Company Minutes, January 30, 1923, March 27, 1923; Charles Deere Velie and 44 Plow in April 24, 1923; Crampton letter in December 21, 1923;

Clausen letter and board response in Deere & Company Minutes, January 29, 1924.

12. Brown diary, April 20, 1923; ibid., May 1, 1923.
13. Deere & Company Bulletin Number 544, January 4, 1924.
14. Model D production for 1925 in Deere & Company Bulletin Number 567, September 11, 1924. Deere & Company Minutes, July 29, 1924, 3896.
15. Letourneau, Peter. *Vintage Case Tractors.* Stillwater, MN: Town Square Books, 1997; Case sold more than 65,000 tractors from 1912–1927, some variations on the same model. Models include the Model 30-60 (1912–1916), 20-40 (1912–1919), Model 40-80 (1913–??), Model 12-25 (1913–1918), Model 102- (1915–1918), Model 9-18 (1917–1918), Model 9-18B (1918–1922), Model 10-18 (1918–1922), Model 12020 (1922–1928), Model 15-27 (1919–1927), and Model 18-32 (1919–1927).
16. Leon Clausen appointment at Case in *The Waunakee Tribune*, July 3, 1924.
17. *Farm Implement News: The Tractor and Truck Review*, July 10, 1924, 9, says he was in the national guard during the "troubles with Mexico"; *Implement & Hardware Trade Journal*, November 24, 1928, 29, reported that he "enlisted in the aviation service of the United States Army and served along the Mexican border. Injuries received there forced his return to civil life."
18. *Fortune*, August 1936.

22 / "Power Farmer"

1. Gray, *Development of the Agricultural Tractor in the United States, Part II, 1920–1950 Inclusive*, 13. Department of Commerce quote in Gray, 15.
2. University of Nebraska Agricultural Engineering Department, Copy of Report of Official Tractor Test No. 117.
3. *Moline Daily Dispatch*, August 6, 1925; November 13, 1925; *Moline Daily Dispatch*, April 13, 1926.
4. *Notes on the Development of the Farmall Tractor*, prepared at the request of H.P. Doolittle by C.W. Gray, September 15, 1932, in Seyfarth, 109.
5. *Jewell-Record* (Iowa), October 7, 1926.
6. *Isabella County Enterprise*, February 26, 1926.
7. Griswold, Glenn. "Farm Implement Firms' Position Unfavorable Weak Must Drop Out," printed in *Chicago Journal of Commerce*, April 3, 1924, in Scrapbook C, 1921–1924, John Deere Archives. In fact, Deere had done just that. Deere sold its timber holdings in Arkansas in 1924, and closed its St. Louis buggy factory. Peek's statement on the Moline Plow Company in *Farm Implement News*, June 5, 1924.
8. Deere & Company Minutes, October 26, 1926.

9. Deere & Company Minutes, January 25, 1927.
10. Deere & Company Minutes, February 15, 1927.
11. Deere & Company Minutes, March 22, 1927.
12. Deere & Company Minutes, April 26, 1927; in the same meeting Wiman reported projected sales of more than 11,000 model D tractors, Deere's largest volume to date. Production, despite "excessive overtime" in March and April, still left them short of satisfying orders. By late summer, the all-crop tractor was ready for production, despite continued debate over the three-row concept. C.C. Webber thought it should be more flexible, built for use as a two- or three-row machine. It would be used by farmers who had already planted with a two-row horse-drawn planter, he reasoned, "in which case the cultivator will be operated as a two-row proposition." If one wanted to convert to a three-row cultivator, that could be done. But Brown continued to defend the three-row concept. In a roundtable discussion, the board continued to discuss the full product lineup. Despite a long line of power farming equipment, the added stresses and pressures of adapting all tools for tractors, it was still considered important to maintain a "line of tools to be horse-drawn as well as tractor-drawn."
13. Speech before the Thirty-First Annual Convention of the Iowa Implement Dealers' Association, Ames, Iowa, January 5, 1927, excerpts printed in *The Harvester World*, February 2, 1927, Wisconsin Historical Society.
14. *The Harvester World*, June 6, 1928.
15. Harvester's full tractor and tractor implement lineups are in *International Harvester Farm Operating Equipment, McCormick-Deering Line* (Catalog) for 1927, 308–340, at Wisconsin Historical Society.
16. Red Baby announcement in *The Harvester News*, February 2, 1927, Wisconsin Historical Society; *Uniontown Morning Herald*, February 23, 1923.
17. *Notes on the Development of the Farmall Tractor*, prepared at the request of H.P. Doolittle by C.W. Gray, September 15, 1932, in Seyfarth 112–113.
18. *Moline Daily Dispatch*, September 15, 1926. The property in East Moline was located east of First Street and north of Fifteenth Avenue.
19. *Moline Daily Dispatch*, February 24, 1927.
20. *Notes on the Development of the Farmall Tractor*, prepared at the request of H.P. Doolittle by C.W. Gray, September 15, 1932, in Seyfarth 115–116.
21. Advertisement in *St. Louis Daily Live Stock Reporter*, December 2, 1927.

23 / "The Layoff Will Be Brief"

1. Bak, 138.
2. *The Washington C.H. Herald*, May 14, 1927.

3. Report recounted in Bak, 136–137.
4. *Edwardsville* (Illinois) *Intelligencer*, May 26, 1927.
5. *Decatur Herald* (Illinois), August 14, 1927.
6. *The Sunday Star* (Washington, DC), November 27, 1927.
7. *Decatur* (Illinois) *Review*, December 7, 1927.
8. Collaboration with International Harvester is outlined in the Theo Brown Diaries, entries for March 13 through September 29, 1926. "Corn borer destroyer" referenced in entry for August 2, 1926. Newspaper clipping on the corn borer appropriation in Theo Brown diary, February 10, 1927. Expenditures for equipment purchased to combat the European corn borer are detailed in "Report of European Corn-Borer Control Campaign by the United States Department of Agriculture for the Period March 14, 1927, to October 31, 1927, Inclusive."
9. Henry Ford quote about Edsel in Bak, 134; *New Castle* (Pennsylvania) *News*, October 28, 1927.

24 / "The Business of Raising Food"

1. "Tractors on Farms Jan. 1, 1928," *Farm Implement News*, January 5, 1928.
2. Deere & Company Minutes, September 27, 1927; history of Model C development in Minutes, October 25, 1927; Deere's board approved the name change from the Waterloo Gasoline Engine Company to the John Deere Tractor Company in January 1926, Deere & Company Minutes, January 26, 1926.
3. Demonstrations at Deere took place January 9–13, 1928; Minutes, February 20, 1928, 4617.
4. Report of Bruce Lourie, January 9, 1928, reported in Minutes, January 17, 1928; February 2, 1928; Deere & Company Minutes, March 26, 1928, April 24, 1928.
5. Deere & Company Branch House Bulletin No. 645, January 20, 1928.
6. Advertisement for the model D in the *Lawrence* (Kansas) *Daily Journal-World*, June 15, 1928.
7. Fordson advertisement ran in papers across the country, including the *Sidney* (Nebraska) *Telegraph*, March 19, 1927.
8. Sorensen, *My Forty Years with Ford*, 240–241.
9. Butterworth's vice-presidential consideration in Broehl, 495–496.
10. Speeches reported in *Farm Implement News*, July 5, 1928; statistics in "Tractors on Farms Jan. 1, 1928," *Farm Implement News*, January 5, 1928.
11. *Moline Dispatch*, July 31, 1928, September 5, 1928, September 21, 1928.
12. Branch House Bulletin No. 650, June 20, 1928.
13. *Farm Implement News*, June 28, 1928, 14–15.
14. *Ludington Daily News*, August 14, 1928.

15. *London Times*, August 29, 1928; *The American Thresherman*, December 1929; and *Farm Implement News*, January 9, 1930, all cited in Seyfarth, 68–69.
16. Ford, *My Life and Work*, 100.

Epilogue

1. Seyfarth, 69; *Farm Implement News*, vol. 50, no. 8, February 21, 1929.
2. *The 1928 Tractor Field Book.* Chicago, IL: Farm Implement News Co., 1928, 13.
3. *Farm Implement News*, vol. 50, no. 10, March 7, 1929, 40.
4. Tractor Shipments from Factory to Branch House Territories, 1922–1938, John Deere Archives. Deere shipped 23,571 model D tractors in 1929 and 11,557 regular GP and 674 GPWT (Wide Tread) tractors.
5. Seyfarth, 76, 88. The 200,00th McCormick-Deering 10-20 was built on June 4, 1930. A photo marking the occasion is at the Wisconsin Historical Society, Image 7241.

INDEX

Page numbers in italics refer to illustrations

ABOUT THE AUTHOR

Photo by Becca Armstrong Photography

Neil Dahlstrom grew up in East Moline, Illinois, part of the Quad Cities and once known as the farm implement capital of the world. He has worked in museums and archives since high school, studying history and classics at Monmouth College (Illinois) and earning his master's degree in historical administration from Eastern Illinois University. Neil began his professional career in Alexandria, Virginia, in an archive documenting the history of the commercial space industry, then moved to the John Deere archives. Since joining John Deere, he has held a variety of roles in history, records management, research, competitive intelligence, communications, and brand management.

A fortunate by-product of research for *Tractor Wars* was re-engaging with his connection to the farm equipment industry. Neil's paternal grandparents met at Minneapolis-Moline. His father and maternal grandfather worked for Case-IH (International Harvester), and his great-grandfather and great-aunt both worked at John Deere.

Neil has appeared on Book TV, National Geographic, PBS, and the History Channel. He is a certified archivist, past chair of the Business Archives Section of the Society of American Archivists, past chair of the Illinois State

Historical Records Advisory Board, and sits on the Kitchen Cabinet, the Food and Agriculture Advisory Board at the Smithsonian National Museum of American History. He's an avid Chicago Cubs fan and enjoys traveling, spending time outdoors with his family, and attempting to keep his 1971 Volkswagen Super Beetle running. He lives in the Illinois Quad Cities with his wife and son.